Advances in Intelligent Systems and Computing

Volume 1291

The series "Advances in Intelligent Systems and Computing" contains publications on theory, applications, and design methods of Intelligent Systems and Intelligent Computing. Virtually all disciplines such as engineering, natural sciences, computer and information science, ICT, economics, business, e-commerce, environment, healthcare, life science are covered. The list of topics spans all the areas of modern intelligent systems and computing such as: computational intelligence, soft computing including neural networks, fuzzy systems, evolutionary computing and the fusion of these paradigms, social intelligence, ambient intelligence, computational neuroscience, artificial life, virtual worlds and society, cognitive science and systems, Perception and Vision, DNA and immune based systems, self-organizing and adaptive systems, e-Learning and teaching, human-centered and human-centric computing, recommender systems, intelligent control, robotics and mechatronics including human-machine teaming, knowledge-based paradigms, learning paradigms, machine ethics, intelligent data analysis, knowledge management, intelligent agents, intelligent decision making and support, intelligent network security, trust management, interactive entertainment, Web intelligence and multimedia.

The publications within "Advances in Intelligent Systems and Computing" are primarily proceedings of important conferences, symposia and congresses. They cover significant recent developments in the field, both of a foundational and applicable character. An important characteristic feature of the series is the short publication time and world-wide distribution. This permits a rapid and broad dissemination of research results.

Indexed by SCOPUS, DBLP, EI Compendex, INSPEC, WTI Frankfurt eG, zbMATH, Japanese Science and Technology Agency (JST), SCImago.

All books published in the series are submitted for consideration in Web of Science.

More information about this series at http://www.springer.com/series/11156

Endra Joelianto · Arjon Turnip ·
Augie Widyotriatmo
Editors

Cyber Physical, Computer and Automation System

A Study of New Technologies

Springer

Editors
Endra Joelianto
Department Engineering Physics
Institut Teknologi Bandung
Bandung, Jawa Barat, Indonesia

Augie Widyotriatmo
Department Engineering Physics
Institut Teknologi Bandung
Bandung, Jawa Barat, Indonesia

Arjon Turnip
Technical Implementation Unit for
Instrumentation Development
Indonesian Institute of Sciences
Bandung, Jawa Barat, Indonesia

ISSN 2194-5357 ISSN 2194-5365 (electronic)
Advances in Intelligent Systems and Computing
ISBN 978-981-33-4061-9 ISBN 978-981-33-4062-6 (eBook)
https://doi.org/10.1007/978-981-33-4062-6

This Springer imprint is published by the registered company Springer Nature Singapore Pte Ltd.
The registered company address is: 152 Beach Road, #21-01/04 Gateway East, Singapore 189721, Singapore

Preface

The conference book comprises the selected papers presented in the International Conference on Cyber Physical, Computer and Automation Systems (CPCAS) 2019 held at Aston Kuta Hotel & Residence, Kuta, Bali, Indonesia, during November 13–15, 2019, published in the Advances in Intelligent Systems and Computing. The international conference was hosted by the Technology, Informatics, Management, and Engineering Research Support (TIMERS), Indonesia, to disseminate the rapidly developed cyber physical systems.

The Cyber Physical, Computer and Automation System (CPCAS 2019) is an international conference aims at providing a forum and place to meet the needs of education, information, research and service related to technology and engineering system and control system in the cyber physical system for interests of industry, academics, and community that always demands continuous development and the study of new technologies. The international scientific meeting is consolidated to stimulate an environment for academics, researchers, students, and practitioners in presenting, sharing, and transferring new ideas, experiences, invention, and information in the evolving and emerging cyber physical systems.

The book comprises 20 papers recommended by invited reviewers that designate research in autonomous system and control, human–machine interaction and human-related engineering, interconnected things, control systems, robotics, industrial automation, measurement, information technology, data mining and monitoring, decision systems, and network technologies. The papers are organized and listed with short description as follows:

- A. Antonyová, P. Antony, and E. Joelianto present "Management of the Computer-Controlled Hygrothermal Properties in the Room Using the Internet with Ventilator Connection" to gain the mathematical modeling describes the hygrothermal properties in case if the ventilator is turn off and turn on, then compared and evaluated through the curve fitting using the least square methods as well as the numerical differentiation.
- A. W. Fadhilah, D. Novita, Emilliano, and A. W. Fadhilah describe "Design of Smart Security with Face Recognition Method on Internet of Things Using Raspberry Pi" to develop a biometric verification system integrated with Internet of things (IoT) that can monitor the security system status and identity the person who accesses system in real time.
- Z. Ayob and G. Weir present "Is Human Behavior the Real Challenge in Combating Phishing" to identify the significant relation between respondents and their behavior in the broad diversity of one Malaysian organization.
- M. Khoiron, A. Alkaff, and R. Effendie describe "Perseverance of Observability in Hovering Quadcopter Under Sensor Faults" to analyze the quadcopter observability and controllability of the augmented model represented as a linear model by taking linearization around operating states.
- M. Turnip, A. Dharma, Andrian, A. Afriansyah, A. Oktarinoa, and A. Turnip present "Integration of FIR and Butterworth Algorithm for Real-Time Extraction of Recorded ECG Signals" to develop a real-time remote monitoring of cardiac activity by integrating the FIR and Butterworth algorithm for ECG signals extractions.
- H. S. Nida, R. O. Bura, and Abdurahman present "Multi-network Transmission Using Socket Programming to Support Command and Control Systems" to design a multi-network transmission system using socket programming as a communication media. Socket enables communication between one host and another using the same protocol.
- D. Novita, A. Abdurrochman, and A. Sholahuddin consider "Control Prototype of Manipulator Robot for Skin Cancer Therapy" to develop a manipulator robot for skin cancer therapy equipped with a laser system that can disintegrate unwanted tissues on the patient's skin such that it is capable of scanning the whole-body skin and of localizing the skin cancers by driving the manipulator in circular or elliptical skimming.
- B. Rahmat, Y. Via, A. Wasian, I. Purbasari, N. K. Sari, W. Wurjani, S. Bandong, E. Joelianto, and P. Siregar present "Video-Based Container Tracking System Using Deep Learning" to apply the YOLO deep learning in order to detect moving containers. The model is trained using container images, and then the validation and testing process are carried out on the model using other container images.
- H. Safari and K. Mutijarsa consider "Multi-Label Classification Using Problem Transformation Approach and Machine Learning on Text Mining for Multiple Event Detection" to find the best models for event detection from user's tweets into multi-label classification using problem transformation approach and machine learning (ML) techniques.

- Herriyance, P. Sihombing, and R. Rivaldo present "Development of an Automatic Control System for Controlling of Soil pH Using a Microcontroller" to describe the development of an automatic soil pH control system to become an intelligent farmer who can find suitable plants on agricultural land.
- Z. A. Solakha, Salmah, and I. Solekhudin present "Influences of Off-Ramp Volumes on Mean Speed, Based on METANET Model" to develop METANET model based on real situation related to influences of the existence of off-ramp volumes as a part of traffic control.
- H. Y. Sutarto and E. Joelianto present "An Adaptive Queue-Length Estimator Based on Stochastic Hybrid Model" to confirm that the hybrid model together with the particle filter (PF) parameter estimator gives satisfactory results, properly capturing the evolution in time of queue length and traffic flow.
- A. Turnip, D. Esti, G. W. G. Arson, and D. Setiadikarunia present "Electric Wheelchair Controlled-Based EMG with Backpropagation Neural Network Classifier" to discuss the design and implementation of signal processing using artificial neural network (ANN) for classification of motion command brain-controlled wheelchair.
- A. Turnip, D. E. Kusumandari, S. A. Sobana, A. N. Istiqomah, T. Hidayat, S. Iskandar, Y. Nabila, R. Amrina, and P. Madona describe "Effect of Methadone on the Brain Activity in Close Eyes Condition" to design an experiment with rehabilitation patients in order to identify the effect of the drugs on the brain activity in the frontal, central, temporal, and occipital lobes.
- S. Vantika, U. S. Pasaribu, S. W. Indratno, and A. Pancoro develop "Estimating Corn Weight Using Mixed Model with Linear Covariance Function Matrix" to estimate plant breeding value from 500 genotyped individual corn (Zea mays) data randomly taken from the population which consisted of 528 genotyped individual corn data.
- I. Wahyuni, F. L. Wibowo, F. Rahman, W. F. Mahmudy, and A. Iriany present "Design and Implementation of Automatic Weather Station Using MQTT Protocol" to develop automatic weather station (AWS) which has the ability to transmit automatically weather data from measurement using technology Message Queuing Telemetry Transport (MQTT) protocol.
- I. Wahyuni1, P. F. E. Adipraja, W. F. Mahmudy, and A. Iriany present "Rainfall Prediction in Tengger Indonesia: A System Dynamics Approach" to model forecasting rainfall not only uses time series data but also include other supporting variables.
- D. Wijayanto, E. Leksono, and A. Widyotriatmo investigate "Adequacy Assessment of Grid-Connected Nanogrid" to compare one system with another and measure the ups and downs of a system's performance over a time period and to use the evaluation result as the base of further improvement.
- I. N. Yulita, A. Sholahuddin, Emiliano, and I. G. W. Putra present "Electroencephalography-Based Neuromarketing Using Pegasos on Partition Membership Data" to examine neuromarketing by using machine learning-based electroencephalography data from 30 respondents.

- A. A. Yunanto, D. Herumurti, S. Rochimah, and S. Arifiani describe "A Literature Review for Non-player Character Existence in Educational Game" to identify and classify information about the educational game for student's knowledge improvements.

The guest editors would like to acknowledge support from the series editors of Advances in Intelligent Systems and Computing: Prof. Dr. Janusz Kacprzyk for his support and guidance in publishing the book in 2021. The contribution from the diligent authors is gratefully acknowledged. The guest editors appreciate all authors for their cooperation in completing the manuscripts. The editors wish the book will be valuable to readers in enhancing intelligent system and computing, particularly in cyber physical, computer, and automation systems.

Bandung, Indonesia

Endra Joelianto
Arjon Turnip
Augie Widyotriatmo

Contents

About the Editors

Endra Joelianto (MIEEE'01) received his B.Eng. degree in Engineering Physics from Bandung Institute of Technology (ITB), Indonesia, in 1990, and his Ph.D. degree in Engineering from The Australian National University (ANU), Australia, in 2002. He was a Research Assistant in the Instrumentation and Control Laboratory, Department of Engineering Physics, Bandung Institute Technology, Indonesia, from 1990 to 1995. Since 1999, he has been with the Department of Engineering Physics, Bandung Institute of Technology, Bandung, Indonesia, where he is currently an Associate Professor. He has been a Senior Research Fellow in the Centre for Unmanned System Studies (CENTRUMS) since 2007, National Centre for Sustainable Transportation Technology (NCSTT) and Centre for Security and Defence since 2016 at Bandung Institute of Technology, Bandung, Indonesia. He is an Associate Editor of the International Journal of Artificial Intelligence (IJAI), India, since 2008, the Editor-in-Chief of the Journal of Instrumentation and Automation Systems (JIAS), Korea–Indonesia, since 2014, and the advisory Editor of the Journal of Mechatronics, Electrical Power, and Vehicular Technology (JMEV), Indonesia, since 2016. He was an editorial advisory board of the International Journal of Intelligent Unmanned Systems (IJIUS), Emerald 2013-2017. He served as a Guest Editor for the International Journal of Applied Mathematics and Statistics (IJAMAS), International Journal of Imaging and Robotics (IJIR), and Internetworking Indonesia Journal (IIJ). He serves as a reviewer for several reputable international journals of IEEE, Taylor Francis, Elsevier, etc. He was the Chair of the IEEE Indonesia Section Control Systems and Robotics and Automation Joint Chapter Societies (CSS/RAS) in 2012–2017. He is the Chapter Chair Coordinator of the IEEE Indonesia Section (2017–now). He received the best paper award in the 3rd International Conference on Kinematics, Mechanics of Rigid Bodies, and Materials 2015 (MECHKINEMATICS 2015), Bali, Indonesia, and the 3rd International Conference on Robotics, Biomimetics, & Intelligent Computational Systems 2018 (ROBIONETICS 2018), Bandung-Indonesia. His research interest includes hybrid control systems, discrete event systems, computational intelligence, robust control, unmanned systems, intelligent automation, and industrial Internet of things. He edited one book on intelligent unmanned systems (Springer) 2009, published one book on linear quadratic control (ITB-Press) 2017, and published more than 150 research papers.

Arjon Turnip received the B.Eng. and M.Eng. degrees in Engineering Physics from the Bandung Institute of Technology, Indonesia, in 1998 and 2003, respectively, and the Ph.D. degree in Mechanical Engineering from Pusan National University, Busan, Korea, under the World Class University program in 2012. He is currently working at the Technical Implementation Unit for Instrumentation Development, Indonesian Institute of Sciences, Indonesia, as a Research Coordinator. He received Student Travel Grand Award for the best paper from ICROS-SICE International Joint Conference 2009, Certificate of commendation: Superior performance in research and active participation for BK21 program from Korean government 2010, JMST Contribution Award for most citations of JMST papers 2011, Inventor Technology Award from Minister of RISTEKDIKTI 2015, and Bupati Samosir Award for the role and activities of Samosir Development. He is Chairman of the IEEE Indonesia CSS/RAS Joint Chapter. He was an Editor of the Widyariset Journal and the Journal of Mechatronics, Electrical Power, and Vehicular Technology (JMEV). He is the Guest Editor at the Indonesian Internetworking Journal (IIJ) and the International Journal of Artificial Intelligence (IJAI). His research areas are integrated vehicle control, adaptive control, nonlinear systems theory, estimation theory, signal processing, and brain engineering such as brain–computer interface. He published more than 125 research papers.

Augie Widyotriatmo received his bachelor's degree in Engineering Physics and master's degree in Instrumentation and Control from the Institut Teknologi Bandung (ITB), Indonesia. He obtained a Ph.D. degree in Mechanical Engineering from Pusan National University, Korea. He is currently an Assistant Professor in Instrumentation and Control Research Group, Faculty of Industrial Technology, ITB, Indonesia. He has published many technical papers in national and international journals and conference proceedings. He received an outstanding paper award in the 18th International Conference on Control, Automation and Systems (ICCAS 2018), South Korea, best paper awards in the 4th International Conference on Industrial Internet of Things 2018 (ICIIOT 2018), Indonesia, and in the 3rd International Conference on Robotics, Biomimetics, & Intelligent Computational Systems 2018 (ROBIONETICS 2018), Indonesia. He serves as an Associate Editor in the International Journal of Control, Automation, and Systems (IJCAS). He was the Chair of the IEEE Indonesia Section Control Systems and Robotics and Automation Joint Chapter Societies (CSS/RAS) in 2017–2019. His current research includes robotics, material handling vehicles, nonlinear control, energy optimization and automation, biomedical instrumentation, metrology, process automation system, and safety instrumented system. He published more than 90 research papers.

Management of the Computer-Controlled Hygrothermal Properties in the Room Using the Internet with Ventilator Connection

Anna Antonyová, Peter Antony, and Endra Joelianto

Abstract Hygrothermal properties have usually the specific features in every building. The temperature and humidity have a direct influence on the building materials as well as the human well-being and comfort feeling. The energy consumption is also predominated with influence of the climate, building materials, and the type of the building envelope. Another factor which can directly influence the conditions in the room is a ventilator. Hygrothermal properties in connection with the air ventilation have also the direct influence on the health of the building residents. In our experimental work, we used the Internet connection for collecting the data to avoid the external influences on experiment running. The resulting data were analyzed using the mathematical modeling as well as the statistical methods. The characteristics of mathematical modeling describe the hygrothermal properties in case if the ventilator is turned off and turned on. The results are compared and evaluated through the curve fitting using the least square methods as well as the numerical differentiation.

Keywords Hygrothermal properties · Internet · Analysis · Ventilator

1 Introduction

Management of hygrothermal properties as well as maintaining its optimum values are of great importance for its immediate impact on both the quality of materials used in the building and health of its occupants [1]. Reviews of environmental requirements

A. Antonyová
University of Prešov in Prešov, Konštantínova 16, 08001 Prešov, Slovak Republic
e-mail: antonyova@gmail.com

P. Antony
APmikro, 08001 Prešov, Slovak Republic
e-mail: peteantony@gmail.com

E. Joelianto (✉)
Faculty of Industrial Technology, Institut Teknologi Bandung, Bandung 40132, Indonesia
e-mail: ejoel@tf.itb.ac.id

E. Joelianto et al. (eds.), *Cyber Physical, Computer and Automation System*, Advances in Intelligent Systems and Computing 1291,
https://doi.org/10.1007/978-981-33-4062-6_1

consider air quality of major significance and recognized the association between respiratory diseases and hygrothermal air properties. A set of experiments was carried for instance at the Cornell University Animal Science Teaching and Research Center [2] in two calf nurseries where they mounted such ventilation system that permitted control of ammonia concentrations, humidity, and temperature. Another way of using ventilation in agriculture is in a greenhouse. Greenhouse microclimate requires both dehumidification with simultaneous heating and ventilation to keep the climate suitable for plant growing [3]. The numerical model was used to collect experimental data for simulation identical with the greenhouse hygrothermal properties. The numerical methods were also used to model ventilator flow. These studies of Shao-Ting J. Lien and Noor A. Ahmed also suggested that the larger diameter ventilators induce greater mass flow extraction rates [4]. The results of initial data were validated as simulations of different ventilator rotations and wind speeds. The experimental work was conducted at the University of New South Wales. The aerodynamics laboratory enables to examine both the internal and the external flows on the inclined rooftop.

Another way in using system to control the hygrothermal values in the requested interval in agriculture is in a poultry house running. As ambient temperature influences the conditions in the room, opening or closing the windows is a part of the regulation process. System's input button reacts on the results of measuring the humidity and temperature conditions in the poultry house and servomotor opens and close the windows as well as it switches on the DC fan and electric bulb to keep the suitable conditions [5]. The concept of artificial specific heat of air is introduced in order to take into account the latent heat released by internal condensation. Hygrothermal properties were monitored also in broiler house where temperature efficiency was determined as varying system parameters such as indoor relative humidity, temperature, and airflow rate [6].

Joshi and More suggested a real-time temperature controller for application in pharmaceutical industry [7]. Also, when Kolaitis and his colleagues compared qualities of external insulation systems with internal ones they stressed higher risk for water vapor condensation when internal insulation [8]. Their findings are primarily based on analysis of the hygrothermal building performance. Not enough ventilation in combination with humidification especially in walls may cost condensation and mold growth. Lee et al. analyzed hygrothermal performance around the built-in furniture during summer and winter to suggest prevention [9]. Efficiency of heating, ventilation, and air conditioning (HVAC) systems are also directly related to energy consumption. Energy consumption can be affected for instance with application of air-to-air heat recovery fan. Those in the supermarkets in winter were examined and illustrated in the scientific work of Kang et al. [10]. In connection with this, the specific conditions are typical especially in public buildings and workplaces, which have consequences also on well-being and work-related stress. The employees that have unfavorable working conditions at working place perceive individual consequences of work-related stress more negatively [11].

One of the ways to save energy in buildings is to transfer heat and moisture from the exhaust air into the outdoor fresh air. The effective tool for that is the usage of energy recovery ventilator. It has also been experimentally performed that enthalpy efficiency of enthalpy exchanger mainly depends on sensible heat efficiency in winter and latent heat efficiency in summer [12]. Wang and Fukuda introduce the Japanese Institute for Building Environment and Energy Conservation that was established in 1999 to set the residential energy efficiency standard to avoid thermal loss. They evaluate the indoor environmental performance with respect to the qualities of the building envelop using environmental simulation program [13].

Carey J. Simonson with his coauthors investigated hygroscopic building materials [14] and their ability to influence the respiratory comfort in conditions of typical Canadian weather. It was proved that moisture storage in hygroscopic building materials together with air ventilation predetermine comfortable indoor conditions. The mini-mum level of heat during the wintertime is set as a regulation with the New York City Department of Housing Preservation and Development as a health in multifamily residential buildings is perceived as a matter of public health policy. Accordingly, the survey on apartment, building, and household was carried to receive information on indoor vapor pressure and household humidifier use [15]. A too dry environment in the buildings is associated with respiratory problems of their residents. On the other hand, too high vapor in connection with some additional specific conditions could cause material and mold problems. A suitable air ventilation system is considered to be one of the important factors for reducing allergens in the air as well as overall improvement in hygrothermal conditions in the building [16]. The importance of humidification has been tested in conjunction with other procedures in patients with respiratory problems while determining the correlation and prediction in their response [17].

While mainly moisture directly affects the maintenance of the quality of the materials inside the building, on the other hand, the materials from which the building is built affect the quality of its air environment [18, 19]. To monitor the quality of air environment in the building room properly means also eliminating the external influences. One way to maintain the conditions in a dwelling without any external effects that can disrupt the interior room conditions with interdependent hydrothermal properties is through a computer-controlled Internet connection. Computer-controlled Internet connection can also be used to organize management processes [20].

2 Methodology

The main task of the research is to ensure optimal distribution of hygrothermal properties in the room to ensure high level of low energy building standard and low level of heat loss. The methodology we decided to use in this respect uses the possibilities of a ventilator. The ventilator was placed in the corner of the room (Fig. 1), for this purpose, also as the problematic humidity is concentrated mostly in the room corners. Placing the ventilator in a corner of the room in this way reduces the risk of increased humidity, condensation phenomena, and the associated conditions for the development of mold manifestation. The correctness of our consideration, in this respect, should be confirmed by reduced humidity in the corner of the room at

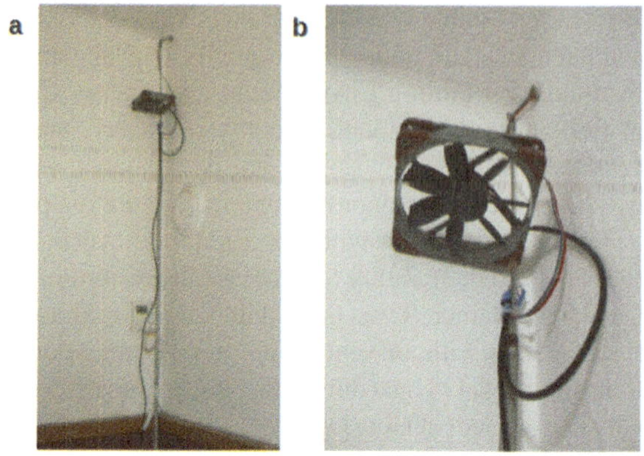

Fig. 1 **a** Ventilator placed in the corner of the room, **b** detail of ventilator with measuring instrument

the ventilator when the ventilator is turned on, as opposed to when the ventilator is not turned off.

Figure 2 comprises detail of the sensor and display with the values, which were obtained during the testing period from October 19, 2018 at 13:30 to October 25, 2018 at 10:30. The values as measuring results were recorded every hour. The ventilator was turned off during the first 24 h. During the next 24 h, the ventilator was turned on. Similarly, the ventilator again was turned off during another 24 h. So, this way the measuring conditions were repeated alternately: turned off and turned on.

Another sensor was placed one meter above the floor level as it is expressed in Fig. 3. This arrangement of the sensors makes it possible to analyze both the heat and humidity distribution in the room with respect to the ventilator. The room was not heated during the measuring period and also there were no outer disturbances as nobody entered the room during the experiment. The temperature as well as humidity values partly depended on the weather. The measurement was realized in the house situated in Central Europe in the municipality that lies at an elevation of 258 meters above the sea level with coordinates 49° 1′ 0″ North and 21° 17′ 0″ East. The room with testing was oriented to the west.

To keep the experimental room conditions without any disturbances that might influence the mutual hygrothermal influences, the measurement equipment as well as ventilator were connected to possibility of Internet control.

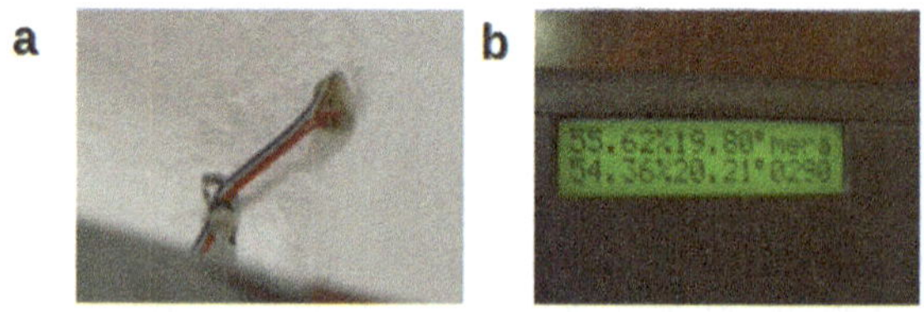

Fig. 2 **a** Measuring instrument that is placed directly at the room corner, **b** measuring results with display

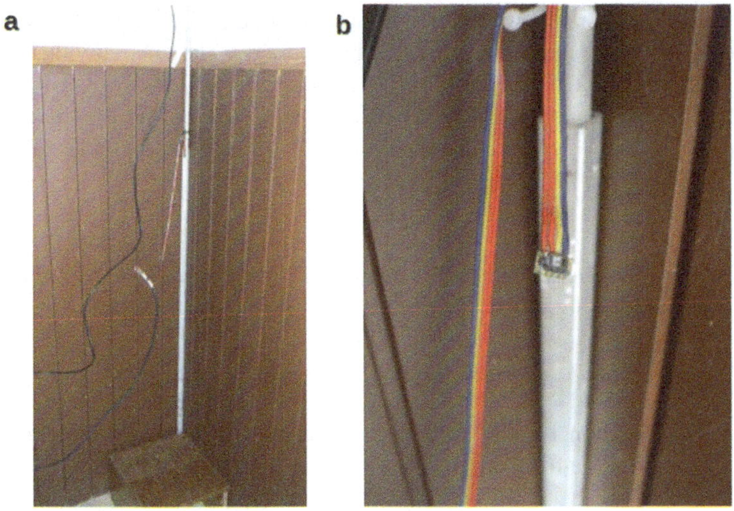

Fig. 3 **a** Measuring instrument that is placed 1 m above the floor level in the experimental room, **b** the detail of the measuring instrument

3 Measurement Equipment Description

There are a few methods how to measure temperature and relative humidity (RH) simultaneously [21]. The measurement in our scientific experimental work was realized through accurate sensors SHT15 and SHT25 from Sensirion manufacturer. These sensors allow the accurate measurement of air temperature with 14-bit resolution and humidity with 12-bit resolution. Position of sensors was determined in relation to the representation of Figs. 1 and 3. Water-resistant coating on the outer parts of protective elements for sensors made of material Super-Isol is there to prevent from seepage of rainwater into the pores of the material. The digital signal from each sensor, SHT15 and SHT25, is processed with a microprocessor AT89S8252 and displayed on the LCD. Said microprocessor also evaluates the time and at the same time every 60 min (3600 s) will register the measured values of humidity, temperature measurement, and serial number to the FRAM memory of FM24C256 with a capacity of 32,768 kB. It is, therefore, performed measurements of 24 entries in 24 h. Each sensor provides two values (temperature + humidity) which were recorded from October 19, 2018 at 13:30 to October 25, 2018 at 10:30.

Both sensors were before the measurement and thereafter calibrated regarding the humidity as well as temperature accuracy. There are a few ways how hygrometers can be calibrated. It is recommended to use slushy mixtures of certain pure salts and distilled water. They have the property to maintain an approximately constant humidity in a closed container. The salts have specific equilibrium humidity levels: lithium chloride ~11%, magnesium chloride ~33%, and potassium sulfate ~97%. We decided to use a saturated sodium chloride bath (H_2O + NaOH) that gives a reading of 75%. Both sensors showed correctly measured humidity value of 75% as it is required. The temperature was controlled by laboratory glass mercury thermometer according to norm STN 258,130.

Signal output requires for sensor SHT15 conversions during programming the microprocessor [22]. Temperature value is for SHT15 given with the formula:

$$T = d_1 + d_2 \cdot \mathrm{SO}_T \tag{1}$$

where d_1 and d_2 are temperature conversion coefficients whose values are for VDD = 5 V when measuring in °C:

$d_1 = -40.1$,
$d_2 = 0.01$

for SOT resolution 14-bit. Temperature compensation of humidity signal is for SHT15 given with the formula:

$$\mathrm{RH}_{\mathrm{true}} = (T_{^\circ\mathrm{C}} - 25) \cdot (t_1 + t_2 \cdot \mathrm{SO}_{\mathrm{RH}}) + \mathrm{RH}_{\mathrm{linear}} \tag{2}$$

where t_1 and t_2 are temperature compensation coefficients whose values are for humidity readout (SORH) 12-bit: $t_1 = 0.01$ and $t_2 = 0.00008$. Relative humidity requires compensation that is given with the formula:

$$\mathrm{RH}_{\mathrm{linear}} = c_1 + c_2 \cdot \mathrm{SO}_{\mathrm{RH}} + c_3 \cdot \mathrm{SO}_{\mathrm{RH}}^2 (\%\mathrm{RH}) \tag{3}$$

where c_1 and c_2 are optimized humidity conversion coefficients whose values are for humidity readout (SORH) 12-bit resolution $c_1 = -2.0468$, $c_2 = 0.0367$, and $c_3 = -1.5955\text{E-}6$, written in semilogarithmic pattern.

Signal outputs for temperature (T) and relative humidity (RH) with result in % require for sensor SHT25 conversions during programming the microprocessor. Sensirion company [22] defines communication protocol with this sensor. SHT25 is set for power 3 V, what must be adapted also in involvement within the scheme. Temperature conversion is for SHT25 given with the formula:

$$T = -46.85 + 175.72 \frac{S_T}{2^{14}} \tag{4}$$

where ST is temperature signal output. Relative humidity conversion is for SHT25 given with the formula:

$$\mathrm{RH} = -6 + 125 \frac{\mathrm{S_{RH}}}{2^{12}} \tag{5}$$

where SRH is relative humidity signal output.

4 Experimental Work

The illustrations to the measurements comprise the figures with ventilator (Fig. 1) and its position in the room corner as the place with the highest humidity concentration. The ventilator has the following parameters:

Dimensions (height x width x depth): 120 × 120 × 25 mm,
Rotation speed (±10%): 700 RPM,
Airflow: 56.9 m^3/h,
Acoustical noise: 6.8 dB/A,
Static pressure: 0.44 mm H_2O,
Input power/input current: 0.60 W/0.05 A,
Voltage: 12 V,
Mean time to failure (MTTF): 150,000 h.

The measuring results are expressed on display (Fig. 2):

- First raw: humidity [%] and temperature [°C] according to the sensor placed 1 m above the floor level,
- Second raw: humidity [%] and temperature [°C] according to the sensor placed in the room corner. The second raw comprises also the measurement order.

The similar way was installed the measuring instrument that is placed 1 m above the floor level as it is expressed in Fig. 3.

5 Results and Discussion the Data Analysis

To analyze the distribution of hydrothermal conditions in the room, mathematical modeling as well as statistical methods were used. The constructed characteristics, which are based on the data, were obtained during the measurements:

- temperature 1 m above the floor level,
- temperature close to the ventilator placed in the corner.

Similarly

- humidity 1 m above the floor level,
- humidity vales close the ventilator placed in the corner.

The differences between the temperatures as well as the differences between the humidity values obtained using both sensors were expressed through the graphical characteristics (Fig. 6) and using the mathematical modeling (6, 7).

In the following, the curve fitting was used to express temperature (9) and humidity (8) that change over time for the third day of measurement from October 21, 2018 at 13:30 to October 22, 2018 at 13:00 during the time, when the ventilator was turned off.

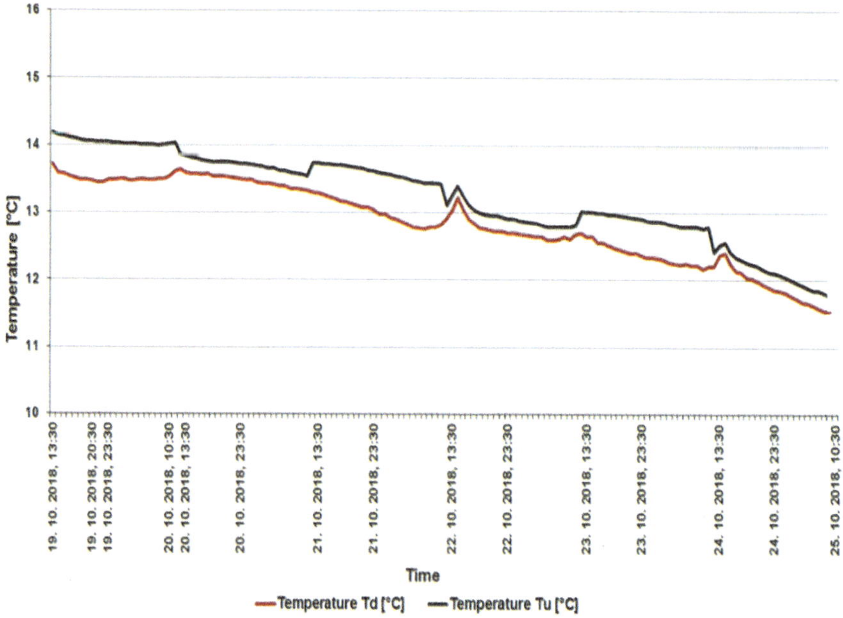

Fig. 4 Graphic characteristics of temperature T_d measured 1 m above the floor level and T_u measured close to the ventilator placed in the upper room corner

5.1 *Temperature*

The temperature values in °C are shown in Fig. 4:

- temperature T_d measured 1 m above the floor level,
- temperature T_u measured close to the ventilator placed in the upper room corner 2700 mm above the floor level.

The measurement results were recorded every hour while the ventilator was alternately turned off and on every 24 h, with the first week turned off. The room was not heated during the measurement.

The graphic characteristic in Fig. 4 shows lower values of the temperature close to ventilator when the ventilator is turned on in comparison to temperature when the ventilator is turned off. The temperature differences are expressed in Fig. 6.

5.2 *Humidity*

Figure 4 shows humidity values in percentage:

- humidity H_d measured 1 m above the floor level,

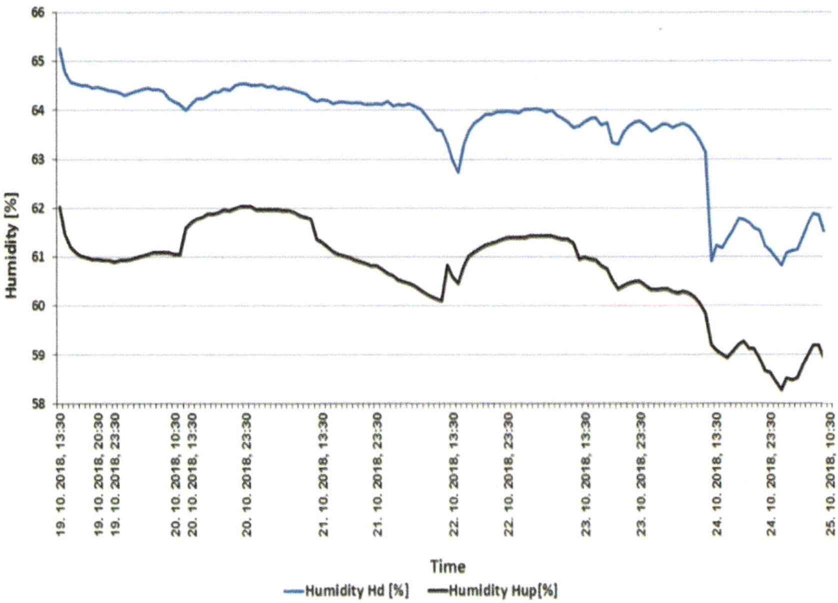

Fig. 5 Graphic characteristics of humidity H_d measured 1 m above the floor level and H_{up} measured close to the ventilator placed in the upper room corner

- humidity H_{up} measured close to the ventilator placed in the upper room corner 2700 mm above the floor level (Fig. 5).

The differences between humidity levels are expressed in Fig. 6.

5.3 Differences

Figure 6 expresses differences between temperatures as well as humidity levels measured 1 m above the floor level and close to the ventilator that is placed in the upper room corner 2700 mm above the floor level. The differences are also expressed (6), (7) using the mathematical modeling while Δt expresses the time period of 1 h between the time when measuring results are recorded, and ti depends on the measuring order.

$$\Delta t = 3600s$$

Difference in temperature:

$$\frac{\partial T(t_{i+1}, z)}{\partial t} \approx \frac{|T_u(t_{i+1}, z) - T_d(t_i, z)|}{\Delta t} \tag{6}$$

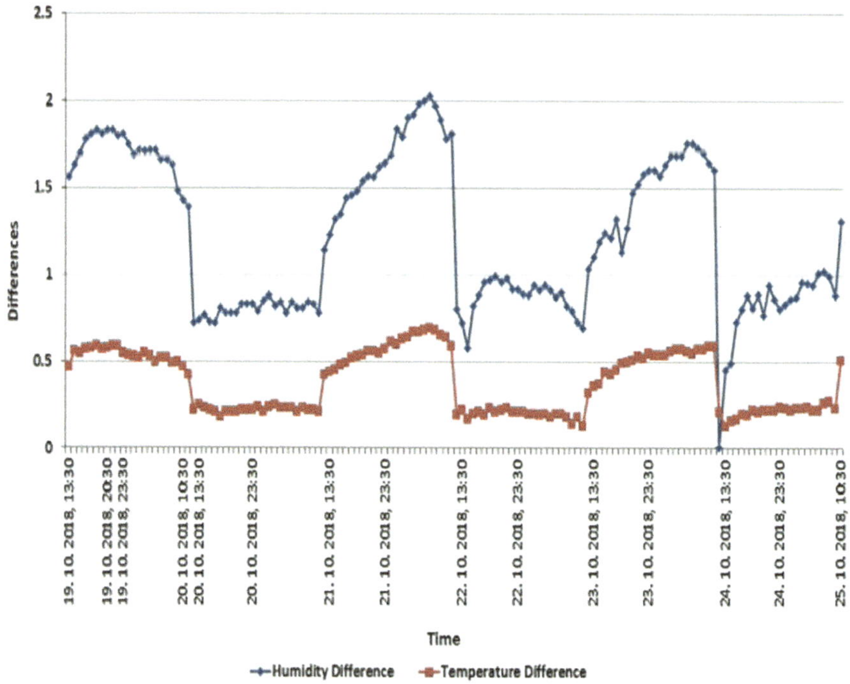

Fig. 6 Differences of temperature as well as humidity values in the time when the ventilator is turned on are lower and more constant

Difference in humidity:

$$\frac{\partial H(t_{i+1}, z)}{\partial t} \approx \frac{\left|H_{\text{up}}(t_{i+1}, z) - H_d(t_i, z)\right|}{\Delta t} \tag{7}$$

The differences of temperature as well as humidity values in the time when the ventilator is turned on are lower and more constant.

5.4 *Curve Fitting*

The differences between temperatures as well as the humidity levels during the third day of measuring process when the ventilator is turned off are not only expressed through graph characteristics in Fig. 7 but also using the statistical method of least squares expressed for curve fitting.

The formula (8) expresses curve fitting using the statistical method of least squares for the humidity while polynomial dependency:

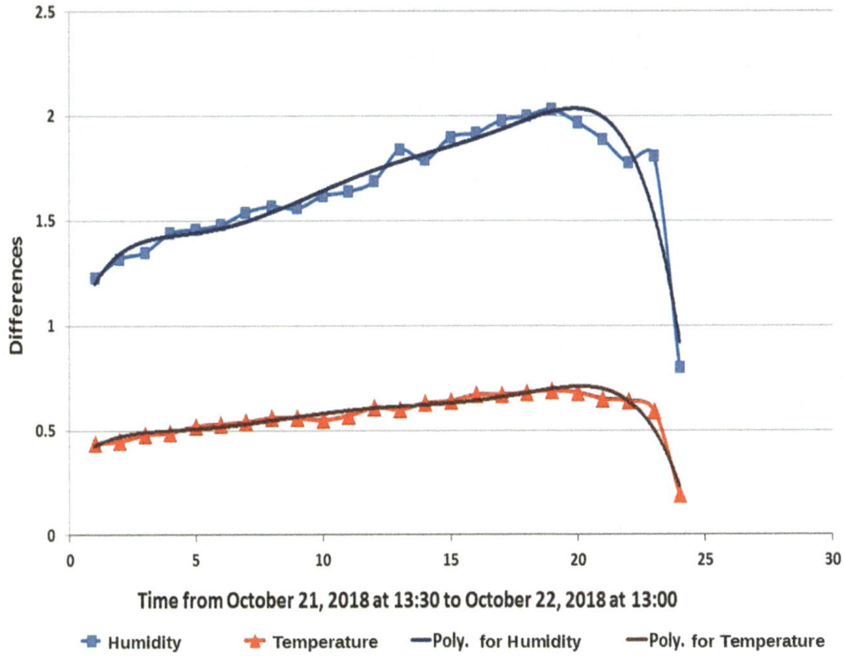

Fig. 7 Curve fitting for the humidity and temperature values while polynomial dependency

$$y = -0.00007x^4 + 0.0028x^3 - 0.0389x^2 + 0.2326x + 0.9863 \tag{8}$$

with the value for index of determination $R^2 = 0.8859$.

The formula (9) expresses curve fitting using the statistical method of least squares for the temperature while polynomial dependency:

$$y = -0.00003x^4 + 0.0012x^3 - 0.0174x^2 + 0.102x + 0.3189 \tag{9}$$

with the value for index of determination $R^2 = 0.8528$.

The values of indexes of determination in both cases, humidity and temperature, proved correct way in expressing the dependency. The computations were realized using the system EXCEL.

6 Contactless Transmission of Signal

The managed contactless transmission of signal can avoid necessity of direct control or setting the measuring equipment, opening the door or the other way influence ambient surrounding of the experimental equipment. Figure 8 comprises schematic

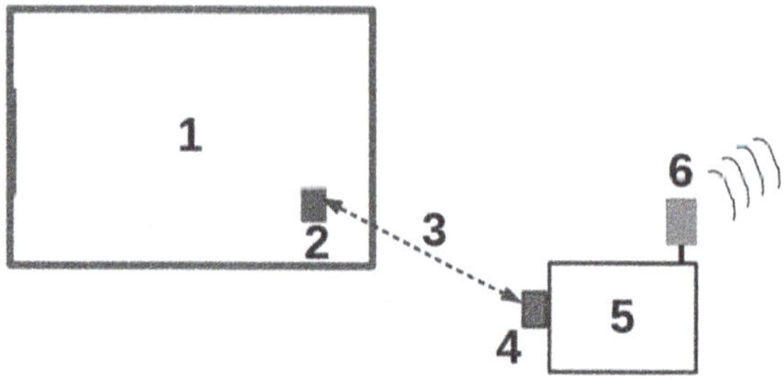

Fig. 8 Schematic expression of the mutual connections and placements of individual equipment

expression of the placements of individual equipment. Part of the equipment is placed directly in the experimental room. Notebook with the data (5) is placed in the independent room. The mutual connection between the rooms is managed as contactless via Internet.

1. experimental room (closed, no interference from door opening),
2. electronic module (measuring temperature, humidity, time, including the data transmission),
3. contactless transmission (signals from serial link of module 2),
4. second electronic module to (2),
5. notebook,
6. module for Internet data transmission.

Figure 9 illustrates the connection of ESP8266 as a complete and self-contained Wi-Fi network that allows contactless transmission and in Fig. 8 it is assigned as 2 in experimental room 1.

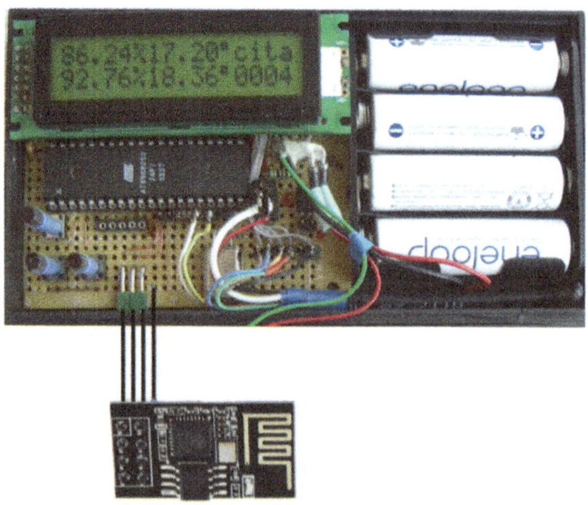

Fig. 9 Connection of ESP8266 as a complete and self-contained Wi-Fi network

7 Efficiency of Ventilator Implication

One of the ways to imagine how effective the ventilator implication in the room corner can be is by comparing the values of the temperature in the state when it is turned off and when it is turned on. Such a quiet ventilator was chosen so that it would not disturb the inhabitants while sleeping. The temperature was measured 24 h without a ventilator running and the next 24 h with the ventilator on, while initially the ventilator was turned off. The dimensions of the experimental room are (height x width x depth): 270 × 350 × 400 cm. The ventilator was placed in the corner of the room opposite the window for gentle air mixing, as the corners of the room have the least air mixing and thus the highest probability of the occurrence of hazardous humidity phenomena.

Temperature in the period (of the year 2018), when the ventilator was turned off:

from October 19, 13:30–October 20, 10:30;
from October 21, 13:30–October 21, 10:30;
from October 23, 13:30–October 24, 10:30.

Temperature in the period (of the year 2018), when ventilator was turned on:

from October 20, 13:30–October 21, 10:30;
from October 22, 13:30–October 23, 10:30;
from October 24, 13:30–October 25, 10:30.

The intervals were determined in this way also because between 10:30 and 13:30 mainly the start-up changes occurred. Figure 10 comprises graphic characteristics, which are composed of points [*i*, *yi*] describing the temperature when the ventilator was turned off and the points [*j*, *yj*] when ventilator was turned on.

$$y_i = \frac{T_i + T_{i+48} + T_{i+96}}{3} \tag{10}$$

where T expresses temperature and $i = 1, \ldots, 20$

$$y_j = \frac{T_{j+24} + T_{j+72} + T_{j+120}}{3} \tag{11}$$

where $j = 1, \ldots, 20$.

The differences $|y_j - y_i|$ are expressed in Fig. 11.

The arithmetic mean of the differences expressed in Fig. 11 is 0.59 °C. If the arithmetic mean of values not affected by start-up changes is expressed, the resulting value is equal to 0.62 °C.

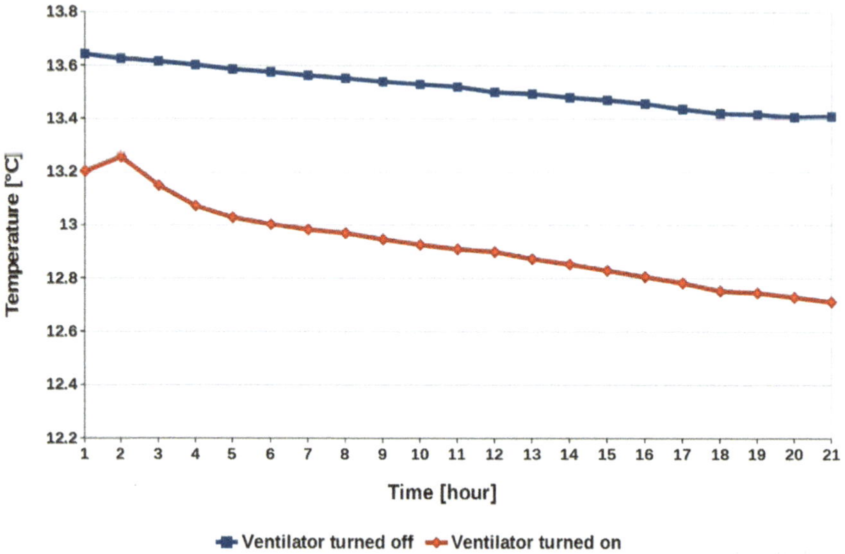

Fig. 10 Graphic characteristics of temperature dependences that are constructed using the formulas (10) and (11)

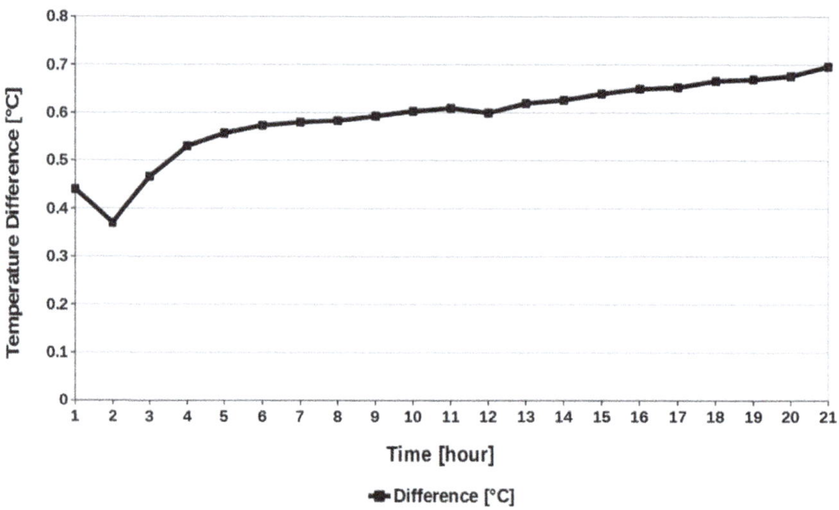

Fig. 11 Graphic characteristics of temperature differences constructed as $|yj - yi|$

8 Conclusions

Optimal distribution of hydrothermal properties in the rooms guarantees high level of low-energy building standard and low level of heat lost. The research focuses the optimal distribution of humidity as well as the temperature in the room including the following hydrothermal performance analysis. The research through the differences between the values of temperature as well as humidity proved that the distribution is more optimal if the ventilator is used. The efficiency of the ventilator application was demonstrated by a temperature difference of 0.62 °C. The ventilator does not disturb as the acoustic noise is of optimum level. The contactless connection to collect the data allowed environment in the experimental room without external influences. Our future research in the field aims to analyze the optimization of the hydrothermal conditions in the room when the room is heated.

Acknowledgements The research was conducted as an integral part of international scientific project with grant 4596-6-17/19 entitled "Modeling of environmental management processes."

References

1. PWE Plant & Works Engineering; Energy Management Homepage: Why is humidity so important? https://pwemag.co.uk/news/fullstory.php/aid/2140/Why_is_humidity_control_so_important_.html. Last accessed 2020/02/20
2. Hillman, P., Gebremedhin, K., Warner, R.: Ventilation system to minimize airborne bacteria, dust, humidity, and ammonia in calf nurseries. J. Diary Sci. **75**(5), 1305–1312 (1992)
3. Kittas, C., Bartzanas, T.: Greenhouse microclimate and dehumidification effectiveness under different ventilator configurations. Build. Environ. **42**(10), 3774–3784 (2007)
4. Lien, S.-T.J., Ahmed, N.A.: Numerical simulation of rooftop ventilator flow. Build. Environ. **45**(8), 1808–1815 (2010)
5. Olaniyan, O.M., Adegboye, M.A., Isife, O.F., Bolaji, O.: Design and Implementation of a temperature and humidity control system for a poultry house prototype. ATBU, J. Sci. Techn. Educ. **6**(1), 106–114 (2018)
6. Nam, S.-H., Han, H.: Computational modeling and experimental validation of heat recovery ventilator under partially wet conditions. Appl. Therm. Eng. **95**(25), 229–235 (2016)
7. Joshi, M.P., More, V.A.: Real time cost effective temperature controller. Int. J. Electron. Commun. Eng. Tech. **3**(2), 271–277 (2012)
8. Kolaitis, D.I., Malliotakis, E., Kontogeorgos, D.A., Mandilaras, I., Katsourinis, D.I., Founti, M.A.: Comparative assessment of internal and external thermal insulation systems for energy efficient retrofitting of residential buildings. Energy Build. **64**, 123–131 (2013)
9. Lee, H.-H., Oh, H.-R., Lim, J.-H., Song, S.-Y.: Evaluation of the thermal environment for condensation and mold problem diagnosis around built-in furniture in Korean apartment buildings during summer and winter. Energy Proced **96**, 601–612 (2016)
10. Kang, Y., Wang, Y., Zhong, K., Liu, J.: Temperature ranges of the application of air-to-air heat recovery ventilator in supermarkets in winter China. Energy Build. **42**(12), 2289–2295 (2010)
11. Kovaľová, J., Frankovský, M., Birknerová, Z., Zbihlejová, L.: Identification of links between sources and consequences of work-related stress. AD ALTA: J. Interdiscip. Res. **8**(2), 65–69 (2018)

12. Liu, J., Li, W., Liu, J., Wang, B.: Efficiency of energy recovery ventilator with various weathers and its energy saving performance in a residential apartment. Energy Build. **42**(1), 43–49 (2010)
13. Wang, Y., Fukuda, H.: the influence of insulation styles on the building energy consumption and indoor thermal comfort of multi-family residences. Sustainability **11**(1), 1–14 (2019)
14. Simonson, C.J., Salonvaara, M., Ojanen, T.: Heat and mass transfer between indoor air and a permeable and hygroscopic building envelope: Part II—verification and numerical studies. J. Therm. Envelope Build Sci. **28**(2), 161–185 (2004)
15. Quinn, A., Shaman, J.: Indoor temperature and humidity in New York City apartments during winter. Sci. Total Environ. **583**(1), 29–35 (2017)
16. Stephen, F.R., McIntyre, D.A., Lane, A.: Ventilation and house air tightness: Effect on indoor temperature and humidity in Southampton, UK. Build. Serv. Eng. Res. Tech. **18**(3), 141–147 (1997)
17. Miller, D.D., Amin, M.M., Palmer, L.B., Shah, A.R., Smaldone, G.C.: Aerosol delivery and modern mechanical ventilation, in vitro/in vivo evaluation. Am. J. Respir. Critic. Care Med. **168**(10), 1205–1209 (2003)
18. Antonyová, A., Antony, P., Joelianto, E.: Modeling of dynamic responses in building insulation. J. Eng. Tech. Sci. **47**(5), 536–548 (2015)
19. Antonyová, A., Antony, P., Korjenic, A.: Laboratory measurement of thermal distribution throughout the insulation materials using the Peltier module while managing elimination of external influences. Energy Build. **128**(15), 336–348 (2016)
20. Antonyová, A., Antony, P., Joelianto, E.: Management of laboratory work for designing manufacturing phases via the ethernet connection. Internetw. Indonesia J. **9**(1), 9–14 (2017)
21. Sun, H., Zhang, X., Yuan, L., Zhou, L., Qiao, X., Hu, M.: An optical fiber Fabry-Perot interferometer sensor for simultaneous measurement of relative humidity and temperature. IEEE Sens. J. **15**(5), 2891–2897 (2015)
22. https://www.sensirion.com/en/environmental-sensors/humidity-sensors/, last accessed 2020/02/23

Design of Smart Security with Face Recognition Method on Internet of Things Using Raspberry Pi

Aldi Wista Fadhilah, Dessy Novita, Emilliano, and Aldo Wista Fadhilah

Abstract The development of security technology has rapidly increased on various aspects, especially for security that is related to the physical entry. Nowadays, traditional security systems are not yet effective in security requirements. The application of technologies such as pin codes and RFID has weaknesses that allow strangers to access the system. Therefore, this paper developed a biometric verification system to integrate with Internet of things that can be monitored the security system status and identity which accesses system in realtime. To implement the biometric verification system, the device should carry out the process of identifying and recognizing the biometric information which is utilized as face information. Viola–Jones algorithm and local binary pattern histogram methods are applied as face recognition methods on the device. Meanwhile, to integrate security devices and the concept of the Internet of things, device is designed to be able to communicate data using the IEEE 802.11 standard and Raspberry Pi as a processor. The experiment results are obtained as smart security with the concept of a biometric verification system that can be monitored and controlled in real time with the integration of the concept of the Internet of things on the device.

Keywords Smart security · Biometric verification system · Face recognition · Internet of things · Raspberry pi

1 Introduction

The security technology has been applied on various fields, such as security systems at homes, banks, laboratories, and so on. These require a good security system to prevent and to avoid strangers to access the system. Nowadays, the traditional security system has not been effective in providing the safety that still has weakness, such as the using of pin codes, RFID, and so on. However, using this technology has

A. W. Fadhilah · D. Novita (✉) · Emilliano · A. W. Fadhilah
Department of Electrical Engineering, Universitas Padjadjaran, Bandung, Indonesia
e-mail: d.novita@unpad.ac.id

E. Joelianto et al. (eds.), *Cyber Physical, Computer and Automation System*, Advances in Intelligent Systems and Computing 1291,
https://doi.org/10.1007/978-981-33-4062-6_2

limitations, while users must ask for a code that gives others access that understands the code or has a card to access the security system. One simple solution that can overcome this problem is to use biometric information as a security component.

Biometric verification system (BVS) is a system that carries out a verification process with reference to biometric information which is obtained from living things through a measurement or calculation process. Biometric information can be distinguished from the characteristics of the body shape of a person which consists of fingerprints, face, DNA, hand shape, and iris. Other classifications are related to a person's behavior such as writing style and voice [1]. BVS can be applied by biometric information recognition techniques which are compared to the database between online recognition. The using of biometric information for security systems can increase the level of security of a system [2]. It can eliminate conflicts such as theft of ID cards and pin codes that are distributed to strangers.

In addition, the need for a security system that can be monitored and controlled remotely is also a matter that should be resolved when implementing a security system. In some cases, the security system does not provide information about the reported anytime and anyone who accesses the system. Remote control of the security system is utilized to provide access to other parties with the user's permission in certain cases. This is related to the BVS agreement that is applied, and then the accessing of system only belongs to users who have been recorded in the database itself. Internet of things technology can be a solution for remote control of the security system [3–7].

Moreover, the smart security system utilizes face as biometric verification system and the local binary pattern histogram (LBPH) method as face recognition method that is selected based on several reasons. There are good performances, both in terms of speed and accuracy of face recognition. The results of the face recognition process are influenced by various factors, where these factors affect the accuracy of the system used. These factors include facial expressions, lighting conditions, rotated digital images, and physical changes in a person's face due to aging [8]. LBPH has also good results under lighting and poses variations [9], whereas face detection is using Viola–Jones algorithm [10–12].

Furthermore, the objectives of this research are as follows. First, security devices are designed and built by implementing the biometric verification system and connection based on the Internet of things. Second, designing security devices that applied face recognition techniques are using the local binary pattern histogram method and Viola–Jones algorithm. Biometric information is a form of face as security information. In designing this device, Raspberry Pi is used as the main controller. Meanwhile, to communicate with the Internet network is utilized the IEEE 802.11 protocol [13, 14], in which case, the communication module used is included in the Raspberry Pi [15–18]. Camera module is required as data capture devices, and then face recognition by image processing is applied to the security system. Through this research, this device is designed to be a solution that suits the needs of security devices for several areas of life. The main contribution of this paper is built a smart security device on Internet of things.

2 Methodology

In the proposed system, the prototype of security system has control system that can be shown in Fig. 1. The input can be captured from camera, and then it is the process to Raspberry Pi by digital image processing and comparison of histograms with database.

2.1 Digital Image Processing

Digital image processing is the main process in the system designed which is implemented in 3 (three) stages, namely as follows [8–12]:

- To convert image from RGB to grayscale
- Face detection using Viola–Jones algorithm
- Face recognition using the local binary pattern histogram method

The three stages of digital image processing can be shown as a block diagram of image processing in Fig. 2, which are implemented on Raspberry Pi with the software

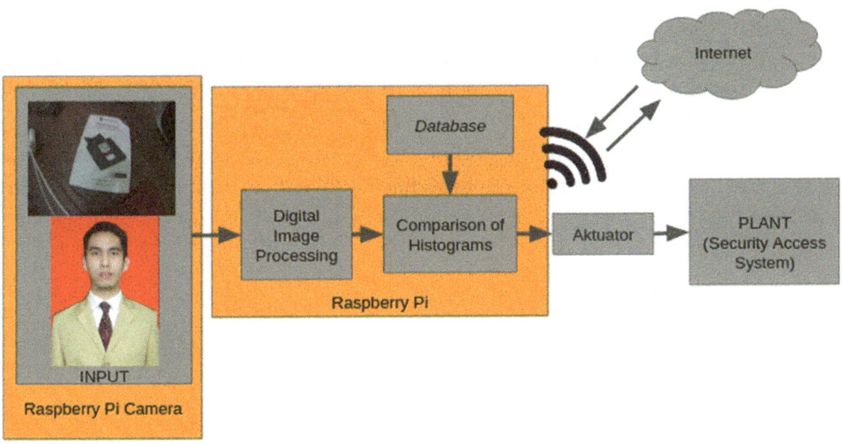

Fig. 1 Block diagram of control system

Fig. 2 Block diagram of digital image processing

created. The software is designed in Python programming language which involves OpenCV as a library [17, 19]. *Viola–Jones algorithm* in face detection process is utilized to detect a human face in digital image input [10–12]. This process is done in *Raspberry Pi* using *Python* as programming language for the software.

2.2 *Internet of Things*

In this research, Internet of things concept is implemented in the security device. The implementation aims to monitor and control the device remotely. IEEE 802.11 Standard is chosen to be utilized as the communication standard in physical and data link layer because of the requirement of the system to send data in the form of image. To communicate through Internet network, the device needs gateway as connecting device to the network [13, 14]. Security prototype consists of power source, camera, actuator, and Raspberry Pi. Data that have been sent to gateway will be transmitted to server. Through this server, users will get the data that have been sent by security prototype. Block diagram of the prototype is shown in Fig. 3.

Telegram app is used as user interface to device. Security device will send data to Telegram bot server so user can monitor and control the device through Telegram. Device will send data as text that tells time and identity who access the system and as image when stranger tries to access the system [20]. The architecture of OSI model of security device is shown in Fig. 4. The architecture IoT has four layers which consist of first layer and is called application layer as Telegram API, the second layer is transport layer as TCP/IP, the third layer is network layer is IPv6, and the fourth layer is combined data link and physical layer as IEEE 802.11 and camera. The proposed system is according to flowchart of face recognition system for smart security in Fig. 5.

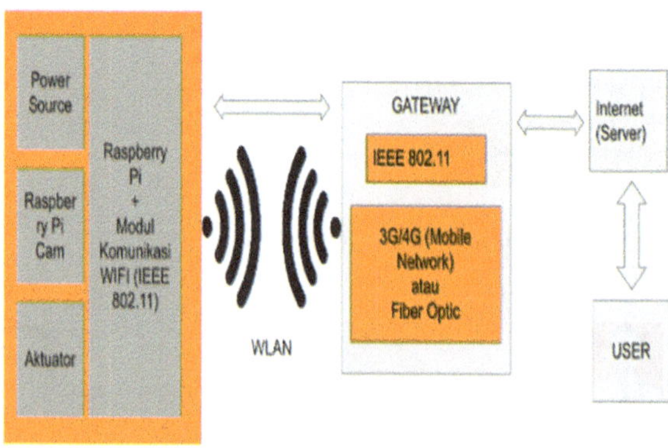

Fig. 3 Block diagram of the prototype

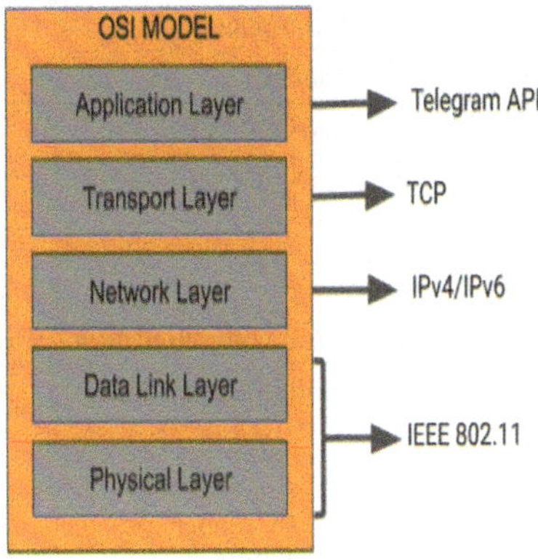

Fig. 4 Data communication architecture (OSI model) of the prototype

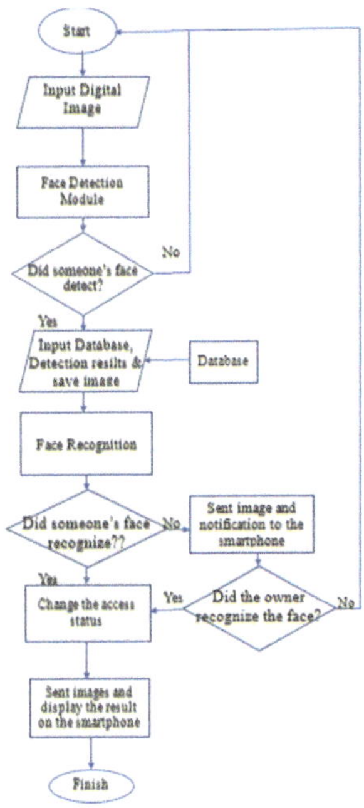

Fig. 5 Flowchart of face recognition system for smart security

3 Experiment and Results

For experiment, the components of prototype that required for proposed system can be described on Table 1. There are Raspberry Pi 3 model B, Raspberry camera, relay, LED, module of communication IEEE 802.11 and smartphone or laptop. The illustration of prototype is shown in Fig. 6 that applied at a room. The position of camera is put on beside the door. To examine the performance of the device, the experiment is held with the setup as follows: (a) The security device that has

Table 1 List of component

No.	Component	Amount	Function
1	Raspberry Pi 3 model B	1	Controller
2	Raspberry Pi Cam	1	Sensor
3	Relay	1	Actuator
4	LED	1	Indicator
5	Module of communication IEEE 802.11	1	Module communication
6	Smartphone or Laptop	1	Display

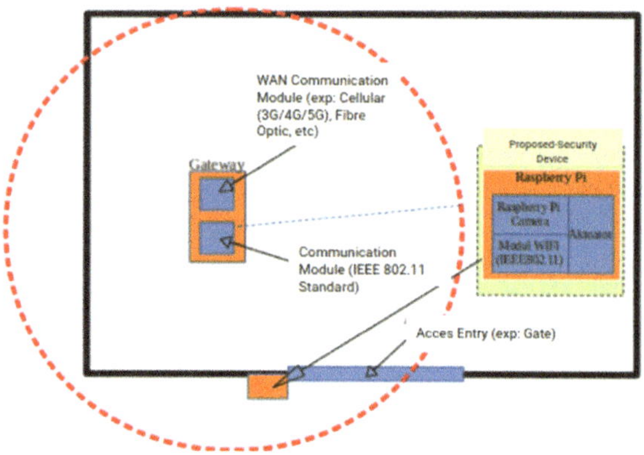

Fig. 6 Illustration of prototype implementation

connected to actuator is turned on and the status of access system is locked. At this moment, Telegram app is also prepared to examine the message that the device sent. (b) Subject A started the experiment process, where subject faced to camera. From this process, the status of the actuator and the message in Telegram is observed to see how good is the performance of the device (This process is repeated 10 times for each subject) (c) After subject A, the process is repeated for subject B and C.

In Fig. 7, the system applied the face detection on the digital image which is taken from *Raspberry Pi camera*. The results of face detection are saved in folder as database. Square symbol is drawn in the digital image for indicating the face that is detected with *Viola–Jones algorithm*. To recognize face, system compares digital images from input device and database in the form of histogram. There is no differences between LBPH method that is used and the original LBPH. The method produced the matching score and provided the label for the identity in the image. The result of this process is shown in Fig. 8.

Figure 8 indicated the identity and the matching score as the results from face recognition process using LBP. The matching score is generated as the result from comparison process between the histogram that is yielded after system operates LBP operation on digital image. The process of comparison between histogram is done by using *Euclidean distance* calculation on Eq. 1.

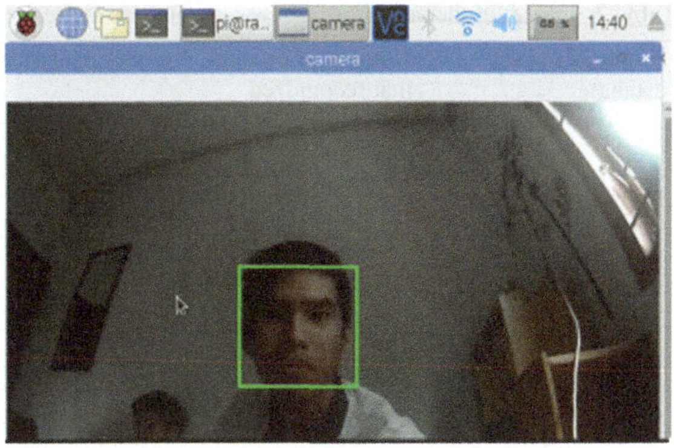

Fig. 7 Face detection's result using Viola–Jones algorithm

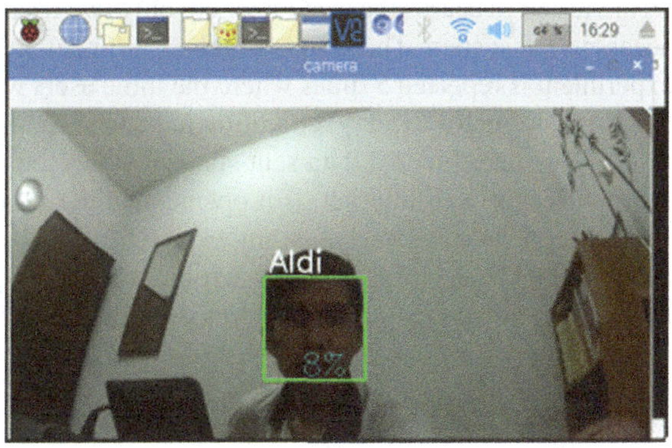

Fig. 8 Face recognition's result by using LBPH method

$$D = \sqrt{\sum_{i,j=1}^{n} \left(\text{hist1}_i - \text{hist2}_j\right)^2} \tag{1}$$

where D is distance (the difference between histogram), hist1_i is histogram that represents database image and hist2_j is histogram that represents input digital image.

If the result of calculation is 0, then the digital image from *raspberry pi camera* is the same as image in the database with 100% matching score. Matching score is declared to percentage and comparing the value with threshold value. The threshold value is 100. To recognize faces, prototype collects user's face image as biometric information and saves it in database, and then it determined the size of memory used and performance of face recognition operation. Table 2 indicated the effect for having various numbers of image in database to the performance of face recognition process. For each image in database, we tested the performance by examining face recognition process at 10 times. The more the images, the slower process the time will be. The proposed security device uses 30 images in database for each user who are given access to the system. The implementation of proposed security device is illustrated in Fig. 6, where the device is located near the access gate and connected to

Table 2 Impact of the number of images in database to face recognition's performance

Number of Images in database	Number of process that recognized successfully from ten experiments	Average process time	Size of memory in used for database (kB)
10 images	4 out of 10 (40%)	0.02745 s	75.7
20 images	10 out of 10 (100%)	0.03352 s	153.1
30 images	10 out of 10 (100%)	0.04377 s	218
40 images	10 out of 10 (100%)	0.05285 s	257.9
50 images	10 out of 10 (100%)	0.0627 s	265.1

gateway via wireless LAN. In the experiment, three users (which labeled as A, B, and C) are tested to access the system where only two users (A and B) have been granted access to the system. The results of the experiment are in Table 3. For each user, the experiment is repeated 5 times where the three users will face to camera and wait for the result. Actuator's (relay) condition indicates whether the actuator is in open state or in close state. Finally, data sent status denotes the status of data that needs to be sent to user through Internet. If it opens, then the status of system is unlocked, and vice versa. System will be only unlocked when device recognized users and relay is active [20]. When stranger enters a room, then smartphone received screenshot of notification on application Telegram that is shown in Fig. 9. The maximum number

Table 3 Results of the experiment

User	Experiment number-N	Face detection result	Face recognition result	Actuator condition (relay)	Data sent status
A	1	Detected	Recognized	Opened	Sent
	2	Detected	Recognized	Opened	Sent
	3	Detected	Recognized	Opened	Sent
	4	Detected	Recognized	Opened	Sent
	5	Detected	Recognized	Opened	Sent
B	1	Detected	Recognized	Opened	Sent
	2	Detected	Recognized	Opened	Sent
	3	Detected	Recognized	Opened	Sent
	4	Detected	Recognized	Opened	Sent
	5	Detected	Recognized	Opened	Sent
C	1	Detected	Not recognized	Closed	Sent
	2	Detected	Not recognized	Closed	Sent
	3	Detected	Not recognized	Closed	Sent
	4	Detected	Not recognized	Closed	Sent
	5	Detected	Not recognized	Closed	Sent

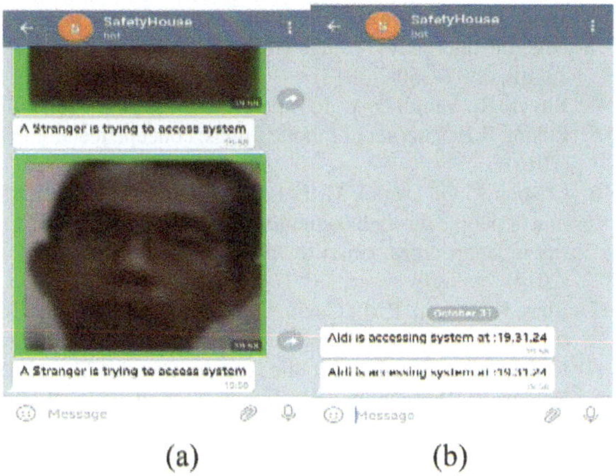

Fig. 9 Screenshot of notification on application Telegram by smartphone, **a** when someone is unknown, **b** known

of granted users that system can handle is not tested in this research but as long as the system can handle the computation, the number of user can be higher than three.

4 Conclusions

Security device that is integrated with Internet of things has been built. The number of images in database was the performance of face recognition's operation. The more images will be the slower process time. According to the result, prototype used 30 images for each user in database in which the average of process time is 0.04377 s. For all conditions, prototype will send data both when the system recognized and not recognized someone, so user of this system can monitor the system remotely.

Acknowledgements The authors would like to express gratitude for Universitas Padjadjaran, Rector of UNPAD, Directorate of Research, Community Service and Innovation of Universitas Padjadjaran, Ministry of Research, Technology and Higher Education of the Republic of Indonesia (RISTEKDIKTI), for research funding of internal UNPAD, Department of Electrical Engineering, Universitas Padjadjaran, fellow research teams and students who participated in this research at Electrotechnic, Electric Power and Communication Technology Laboratory.

References

1. Ross, A.K., Jain A.: Multimodal biometrics: An overview. In: Proceedings of 12th Europian Signal Processing Conference (EUSIPCO), pp. 1221–1224, September (2004)
2. Cetin, G.: Face and recognition on field programmable gate array. In: M.Sc. thesis, Grad. School of Natural and Applied Science, Dokuz Eylul University, Izmir (2010)

3. Wang, C., Daneshmand, M., Dohler, M., Mao, X., Hu, R.Q., Wang, H.: Guest editorial—special issue on Internet of Things (IoT): architecture, protocols and services. In: IEEE Sensors J. 13(10), 3505–3508 (2013)
4. Buyya, R., Vahid, D.A.: IoT: principles and paradigms. Elseiver, Cambridge (2016)
5. Pratim, R.P.: Internet of robotic things: concept, technologies, and challenges. In: IEEE Access (2016)
6. Cristian, G.G., Daniel, M.-L. B., Cristina, P.G.-B., Juan, M.C.L., Nestor, G.-F.: Midgar: detection of people through computer vision in the Internet of Things scenarios to improve the security in smart cities, smart towns, and smart homes. In: Future Generation Computer System (2017)
7. Elias, K., Saraju, P.M., Gavin, C., Umar, A., Prabha, S.: Design of a high-performance system for secure image communication in the internet of things. IEEE Access Special Sect Secur Reliab Aware Syst Des Mobile Comput Dev **4**, 1222–1242 (2016)
8. Rahim, A., Hossain, N.: Face recognition using local binary pattern. Global J. Comput. Sci. Tech. Graph. Vision **3** (2013)
9. Nashwan, A.O., Ilhan, A.: A face recognition method in the internet of things for security applications in smart homes and cities. In: 6th International Istanbul Smart Grids and Cities congress and Fair (ICSG) (2018)
10. Dabhi, M., Pancholi, B.: Face detection system based on Viola-Jones algorithm. Int. J. Sci. Res. **5**(4) April (2016)
11. Viola, P., Jones, M.: Rapid object detection using a boosted cascade of simple features. In: Computer Vision and Pattern Recognition Conference (2001)
12. Parande, A.: Understanding and implementing the Viola-Jones image classification algorithm. https://medium.com/datadriveninvestor/understanding-and-implementing-the-viola-jones-image-classification-algorithm. Last accessed 2019/2/6
13. Ramia, B., Amin, B., Ashraf, A.: A comparison between IEEE802.1a, b, g, n and ac standards. IOSR J. Comput. Eng. **17**(5)
14. Frenzel, L.: 12 Wireless options for IoT/M2M: diversity or dilemma? https://www.electronicdesign.com/iot/12-wireless-options-iotm2m-diversity-or-dilemma. Last accessed 2019/2/6
15. Pajankar, A.: Raspberry Pi Image Processing Programming. Apress, Nashik (2017)
16. Raspberry Team: Raspberry Pi 3 Model B. https://thepihut.com/products/raspberry-pi-3-model-b. Last accessed 2019/2/6
17. Cox, T.: Raspberry Pi Cookbook for Phyton Programmers. Packt Publishing Ltd, Birmingham (2014)
18. Novosel, R., Meden, B., Emeřsič, Z., Štruc, V., Peer, P.: Face recognition with Raspberry Pi for IoT environments. In: Conference paper, September, (2017)
19. OpenCV Team: About OpenCV. https://www.opencv.org/about.html. Accessed 2019/2/7. Williams, R.: What is Telegram: the New WhatsApp? https://www.telegraph.co.uk/technology/news/10658647/What-is-Telegram-the-new-WhatsApp.html. Last accessed 2019/2/9
20. Bolton, W.: Programmable Logic Controller. Burlington, Elseiver (2006)

Is Human Behavior the Real Challenge in Combating Phishing

Zalina Ayob and George R. S. Weir

Abstract Computers can enhance our work activities, e.g., through greater efficiency in document production and through ease of communication. Although reliance on e-mail has reduced with the introduction of instant messaging applications, it continues to hold its own as a preeminent Internet-based communication service. Along with this eminence, we have persistent issues arising from e-mail-borne malware, phishing, and embedded malicious Web links. Existing steps to protect e-mail users still fail to address a significant proportion of online threats every year. A widely held view that we endorse is that this continuing challenge of e-mail is not wholly technical in nature and thereby cannot be entirely resolved through technical measures. Rather, we have here a socio-technical problem whose resolution requires attention to both technical issues and the specific attitudes and behavioral characteristics of end users. In this study, a structured questionnaire was used to collect data from 181 respondents while an experiment was used to identify the significant relation between respondents and their behavior in the broad diversity of one Malaysian organization.

Keywords Phishing e-mail · Human characteristics

1 Introduction

In today's world, communicating via e-mail has become an integral part of daily activity. The most used communication medium nowadays is e-mail. This is not restricted to professional or academic settings but is also engaged in informal environments. In fact, e-mail is more popular than memos or bulletin boards within companies and diverse organizations. Even people in the same house may send documents and news via e-mail in preference to other messaging systems that may have limitations on document types and size. The surprisingly fast acceptance of e-mail is

Z. Ayob · G. R. S. Weir (✉)
Department of Computer and Information Sciences, University of Strathclyde, Glasgow, UK
e-mail: george.weir@strath.ac.uk

E. Joelianto et al. (eds.), *Cyber Physical, Computer and Automation System*, Advances in Intelligent Systems and Computing 1291,
https://doi.org/10.1007/978-981-33-4062-6_3

evident from the executive summary of the E-mail Statistic Report, 2013–2017 [1]. The number of e-mail accounts worldwide is expected to grow 6% annually over the next four years from 3.9 billion in 2013 to over 4.9 billion accounts by the end of 2017. Meanwhile, business e-mail accounts expect an annual growth of 5% annually to reach over 1.1 billion in the same timescale.

Naturally, measures have been developed to reduce the negative impacts associated with the use of e-mail. Spam filters are deployed to detect suspicious content and may automatically remove attachments that are known to be risky, such as executable files and scripts that might be Trojan malware. To circumvent such measures, attackers may seek to hide their activities from users and make e-mails appear legitimate. Understanding the various types of e-mail attack technique is essential for combating attacks with specific purposes especially stealing user's sensitive information such as account numbers, login IDs, passwords, and credit card numbers.

Often, these attacks are difficult to detect using conventional, signature-based, and sender verification techniques due to new 'phishing' style attacks that involve advanced social engineering techniques [2]. Moreover, the phishing attacker is no longer focusing only on stealing recipient data and spreading viruses but extending to high impact data breaches and ransomware.

According to the Wombat 2016 State of Phish report, 10 s of millions of phishing e-mails were sent in the 12-month period between 2015 and 2016, which is a 155% increase from their last report [3]. Spear phishing e-mails, for instance, may target specifically by name, position, and division in an organization instead of relying on generic titles as in broader phishing campaigns or 'Advanced Targeted Attacks.'

As a developing country, Malaysia aims to serve its citizens with advanced technology by using Internet communication in all government services. This is to provide better services to the citizens and develop efficient administration in government. Despite spending millions in investment into advance technology services, migration threat defense facilities and software, the reputation of companies is still at risk if they fail to defend from 'inside.'

Knowing your staff characters is the best way to manage the insider threat. Therefore, this study aims to determine the factors that influence staff in one Malaysian company to respond to phishing e-mails.

2 Theoretical Background

Technology acceptance is about how people accept and adapt the use of technology in their daily life [4]. The success or failure of one technology can be viewed in terms of user engagement with technology use. In similar vein, Iahad and Rahim [5] characterize the value of a technology to its acceptance and uses.

E-mail is a technology that is broadly used to increase user productivity and enhance communication. Therefore, user acceptance of this technology is not doubted today. Extensive use of this technology in everyday activities however has also increased the abuse of associated systems [6]. As a result, researchers interested in studying issues related to technology acceptance focus on individual user characteristics, such as cognitive style, internal beliefs, and their impact on usage behavior [7].

2.1 Unified Theories of Acceptance and Use of Technology Model (UTAUT)

The Unified Theories Of Acceptance and Use of Technology (UTAUT) model proposed by Vankatesh [8] integrates eight elements from other prominent models, namely Theory of Reasoned Action (TRA) [4], Motivational Model (MM) [9, 10], Theory of Plan Behavior (TPB) [11], Theory Acceptance Model (TAM) [9, 12], Innovation Diffusion Theory (IDT) [13], Social Cognitive Theory (SCT) [14], Motivational Model (MM) [10], and Personal Computer Utilization (MPCU) [15]. Vankatesh claimed that UTAUT can explain 70% of technology acceptance behavior and affords better explanation for factors influencing the individual's intention and usage. UTAUT consists of four core determinants of intention to use: performance expectancy, effort expectancy, social influence, and facilitating conditions. Other variables used in the model are gender, age, experience, and voluntariness of use. Expanding by adding extension to ATUT has been applied to the study of a variety of technology [16, 17] in both organizational or non-organizational contexts. Over 400 articles in our concern cited the original UTAUT as their reference to work with some of the extended and modifying factors. The fundamental UTAUT model was demonstrated in Fig. 1. Adding factors according to study in support of investigation of theories may reveal the breakdown of theories results, create new knowledge and generalizability of UTAUT [18].

2.2 Usage

Four items related to usage have been asked to measure the significant correlation between the items with response to phishing e-mail (Table 1). Our hypothesis in relation to usage is, *user who works with e-mail every day will easily detect the legitimate e-mail and not respond to phishing e-mail* [19]. We found the number of e-mail in users' account contributes to the ability on making comparison between previously obtained information with the information from phishing e-mail. However, experience alone is not relevant to measures of how users experience and used their e-mail. Users may receive an e-mail that is not related to their daily activity. A specific questionnaire was designed to find the intention of users in using their e-mail.

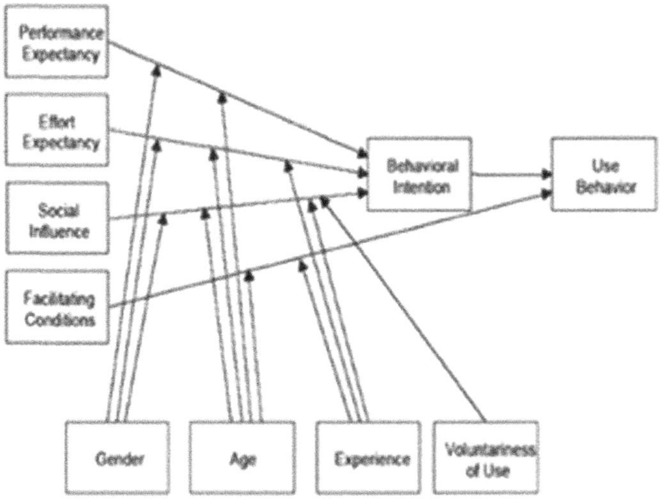

Fig. 1 UTAUT model. *Source* Venkatesh et al. [18]

Table 1 Usage items

Code	Items
US1	I use e-mail to keep up-to-date with work activities
US2	I use e-mail for my social network activities
US3	I use e-mail to pass the time
US4	I found e-mail is easy to use

2.3 Manage

In addition, to UTAUT fundamental constructs, we include 'Manage' as an additional subconstruct to the model. We asked six items under the 'manage' variable shown in Table 2. Our Manage variable objectives are to know how the user manages their e-mail.

Table 2 Manage items

Code	Items
MN1	I manage to control incoming e-mail myself
MN2	I have no problem to tell others my e-mail address
MN3	My e-mail use if effective
MN4	I always monitor my incoming e-mail
MN5	I always monitor my incoming e-mail
MN6	I delete e-mail from unknown sender

Two main types of e-mail management strategies relate to different types of 'refinding' behaviors [20, 21]. The first management strategy is preparatory organization, where users create and tag folders manually with the objective of ease of retrieval. These conventional methods have been replaced with auto-tagging by Gmail. This contrasts with another strategy and opportunistic management which require user's effort in refinding behaviors such as scrolling, sorting, and searching which may compromise user productivity. However, none of these methods is perfectly matched to all users without examining how users manage and access e-mail especially when they are dealing daily with a huge number of inbox e-mails . This may lead users to wrongly monitor and manage their e-mail. We therefore examine the users' e-mail management by giving them a short and simple question asking about their action when dealing with e-mail. See Table 2.

2.4 Elaboration of Likelihood Model (ELM)

The elaboration likelihood model (ELM) aims to explain the behavior theory and users' response to e-mail based on a two-process model: (a) the central route processes where decisions are made based on examining the content, which comes from comparison to prior belief; (b) judging the message based on the appearance of the message, called peripheral route process [22]. Appearance here is likely to include how the message was presented in terms of its font, color, image, and structure. All such features may serve to trick the user's visual sense and thereby trick unaware or agreeable types of user to respond to phishing e-mail. Users who rely on the peripheral route are particularly inclined to respond to phishing e-mails.

2.5 Trust

There is no specific definition of trust to date although many research publications include trust as an important element of study [23–25].

As Internet users, trust on the Internet refers to an 'individual's perceptions of the institutional environment, including the structures and regulations that make an environment feel safe' [26]. Also, according to Carter and Bélange [27] trustworthiness is the main factor that influences users to use e-government services. In this research, trust is used to measure the direction of user action. The drawback of relying on trust is that victims are unable to distinguish the legitimacy of the e-mail. Trust was measured using five modified items based on McKnight et al. [26]. The five items used to measure the trust variable in the model are shown in Table 3.

Table 3 Trust items

Code	Items
TR1	I found e-mail is a good way to maintain relationship
TR2	I have no problem to tell others my e-mail address
TR3	My e-mail use if effective
TR4	I always monitor my incoming e-mail
TR5	I only response e-mail from known sender

2.6 Big Five Personality

The Big Five theory of personality accounts for individual characteristics in terms of five dimensions: (a) extroversion, (b) agreeableness, (c) consciousness, (d) openness, and (e) emotional stability. Several studies [28, 29] have sought to validate the overall structure integrity between these traits and human behavior in a variety of different settings. We chose to use these five traits in relation to responses to phishing e-mail after considering two questions:

i. Are the Big Five personality traits related to vulnerability and response to phishing e-mail?
ii. Which from the five personality traits have most influence in the user's response to phishing e-mail?

3 Research Model

The process of identifying the characteristics to observe can be crucial. We must consider interpersonal characteristics, skills, habits, and even the change of culture and inclination to accept technology. Figure 2 shows the theoretical framework deployed in this research. The unified theory of acceptance and use of technology model (UTAUT), and the elaboration likelihood model (ELM) were adopted as fundamental frameworks in the present study with the extension of trust and Big Five personality elements. We also enhanced our model by including demographic factors to embrace culture differences and to capture the influence of user background and inclination to respond to phishing e-mail.

In this study, surveys were distributed, using e-mail in English, to employees in the Malaysian government sector under the Ministry of Rural and Regional Development. From the 1500 invited participants, only 190 surveys were returned, giving a response rate is 12.6%. Of the returned surveys, 181 were considered valid for further analysis where the others were rejected as incomplete . However, this level of response still falls within the accepted range for use with the structural equation modeling technique [30].

The statistical software package SPSS (version 16.0) was used to estimate the reliability of the questionnaire and identify the intercorrelation value between variables. Regression analysis was applied to the satisfied observable variable. In this

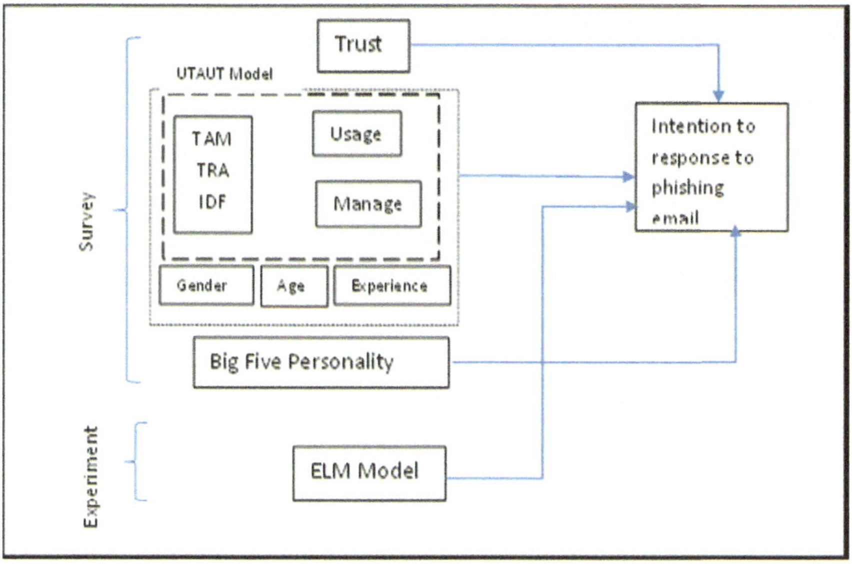

Fig. 2 Theoretical framework

paper, we limit discussion to our result from SPSS that involves investigating the impact of various predictors on singular dependent variables [31].

To support the objective of identifying which independent variables bring the user to respond to phishing e-mail, we use linear and logistics regression in SPSS. The advantage of linear regression is the ability to test the regression coefficient individually between variables while logistic regression is more specific for use on binary categorical dependent variables.

The 'Response' variable was recorded using binary input. Three e-mail experiments were set up to track 'user clicks' on different types of fake phishing e-mails. The experiment used e-mail 'open' detection software which is designed to support our Microsoft Outlook 365 e-mail client and Google analytic software for link click and Web open tracking. Data recorded from analytic software was then matched with logs provided by the organization. Only users who responded to both the phishing e-mail and the survey were marked as a response and were used as binary input for logistic regression.

4 Results

Notably, the response rate from female respondents was almost two times that of males. This is not surprising since the experiment targeted the headquarters where more than 50% of operation was clerical and office-based work which is mainly populated by females. Also, the majority of respondents were between the ages of

Table 4 Respondents' e-mail usage priority

Usage		Frequency	Percentage
a. Number of e-mail account	1 Only	25	13.8
	2–4 Account	150	82.9
	>5 Account	6	3.3
b. E-mail service	Gmail	64	35.4
	Company e-mail/outlook	106	58.6
	Yahoo	5	2.8
	Others	6	3.3
c. Priority of e-mail	Work communication	94	51.9
	Online verification	46	25.4
	Online shopping and promotion updates	21	11.6
	Social networking notification	20	

30 and 35 years old (28.1% of all respondents). In addition, 58% of the respondents were from top-level or executive management.

4.1 E-Mail Usage and Practice

Among the respondents, 69.1% of the total have more than 10 years' experience in using e-mail, with 21% having 5–10 years' experience. From the respondents with more than 10 years' experience, 41% reported having more than 500 e-mails in their inbox, and 24.3% had 101–500 e-mails in their inbox. Table 4 shows the e-mail service distribution and the number of users' e-mail accounts. The study also revealed that 'use of e-mail for work communication' was the primary purpose.

4.2 Variable Validity

Principal component analysis (PCA) was used for factor analysis to determine the validity of item constructs. Factor analysis was used to reduce the large number of variables to a smaller set. Although in this experiment the number of variables involved was not large, identifying the variable(s) that contribute most to the analysis was our objective.

A reliability test and factor loading were performed with Cronbach's alpha test [32] to measure internal consistency between items (see Table 5).

The minimum requirement for factor loading (>0.4) for exploratory research is met by the items in the survey. Table 5 indicates those items that are accepted and rejected (drop) for further analysis in the experiment.

Table 5 Variables factor loading and reliability analysis

Variables	Items	Alpha	Loading factor	Item drop
Trust	TR1	0.685	0.924	
	TR2		0.851	
	TR3		0.931	
	TR4		0.435	Drop
	TR5		0.587	Drop
Usage	US1	0.540	0.685	
	US2		0.853	
	US3		0.618	
	US4		0.312	Drop
Manage	MN1	0.525	0.698	
	MN2		0.875	
	MN3		0.839	
	MN4		0.878	
	MN5		0.605	
	MN6		0.513	Drop
Vulnerability	EM1	0.688	0.595	
	EM2		0.814	
	EM3		0.688	
	EM4		0.779	

4.3 Variable and Hypothesis Testing

This research involved two dependent variables: vulnerability and response. Therefore, the experiment was done using linear regression and logistic regression to investigate the impact of unique variables on their dependent variable. The result for linear and logistic regression and its summary are shown in Tables 6, 7, 8 and 9, respectively.

Table 6 Linear regression with vulnerability as dependent variable

Model	Unstandardized coefficients		Standard coefficients	*t*	Sig
	B	Std. error	Beta		
Constant	−2.423	1.577		−1.536	0.127
Usage	0.640	0.223	0.278	2.871	0.005
Agreeableness	0.278	0.124	0.205	2.236	0.027
Conscientious	−2.52	0.108	−0.194	−2.325	0.022

Table 7 Model summary—vulnerability

Model	*R*	*R* square	Adjusted *R* square	Std. error of estimation
1	0.496	0.246	0.185	1.034

Table 8 Logistic regression with respond as dependent variable

Model	*B*	S.E	Wald	*df*	Sig.	Exp (*B*)
Trust	0.165	0.293	4.316	1	0.049	1.179
Manage	−0.740	0.397	3.471	1	0.050	0.477
Extrovert	0.399	0.241	2.726	1	0.037	1.490
Vulnerability	−1.765	0.497	12.629	1	0.000	0.171

Table 9 Model summary—respond

Step	−2 log likelihood	Cox and Snell *R* square	Nagelkerke *R* square
1	116.398	0.189	0.274

5 Discussion

This experiment was carried out to validate the relationship of independent variables with identified behavior factors that influence user response to phishing e-mail in our model. From the results, it can be seen that users' characteristics are able to influence their behavior toward responding to phishing e-mail and becoming victims.

Correlation analysis between variables starts with identifying the reliability of each variable construct and continues with regression. Two types of regression were used to support the different types of dependent variables' data.

The results show that users high in agreeableness and usage are more likely to respond to phishing e-mail, while conscientiousness in personality does not have direct influence on the level of vulnerability. Logistic regression shows that only users with high extrovert and trust personality have a tendency to respond to phishing e-mail.

Based on these results, two characteristics from Big Five personality have a significant effect on user behavior to become vulnerable and respond to phishing e-mail.

Phishing e-mails common trends require victims to respond and comply with their requests, which may be the reason that extroversion and trust significantly impact users' response to phishing e-mails. A lack of suspicion is unable to tell the victims the difference between phishing e-mails and legitimate ones.

Our finding is supported by other research [33, 34] which found that the same characteristics, extroversion and conscientiousness from the Big Five personality traits, reflect user behavior in respect of Internet usage.

Based on our research, we conclude that users' characteristics contribute to the capability of phishing e-mails detection. In organizations, management plays an important role in planning the appropriate training to improve user behavior against business threats that come from inside.

Therefore, knowing the type of staff you have can give an idea of (a) what type of education is most important to improve user detection capability and (b) whether your company requires reassistance strategies for users' risk-taking behavior .

6 Limitation

Our research is limited to participants from one organization and with a limited number of participants. The results cannot be generalized to the wider population and other organizations that might have a proper threat awareness program directed to end users. To generate broader results, similar experiments must be conducted for the enormous number of participants in different working environments and organizations.

References

1. Radicati, S., Analyst, P., Levenstein, J.: Email Statistics Report, 2013–2017 **44**(0), 2013–2017 (2013)
2. Aaron, G., Rasmussen, R.: Global phishing survey : trends and domain name use in 2H2013. Apwg, no. April, pp. 1–31 (2014)
3. Wombat, S.: A Wombat Security Research Report, January 2016 (2017)
4. Louho, R., Kallioja, M., Oittinen, P.: Factors affecting the use of hybrid media applications. Graph. Arts Finl. **35**(3), 11–21 (2006)
5. Iahad, A., Rahim, A.: A comparative study of acceptance and use of ICT among university academic staff of ADSU and LASU: Nigeria **2**(1), 103–115 (2012)
6. de Paula, R., et al.: In the eye of the beholder: a visualization-based approach to information system security. Int. J. Hum Comput Stud. **63**(1–2), 5–24 (2005)
7. Dillon, A.: User Acceptance of Information Technology (2001)
8. Morris, M.G., Davis, G.B., Davis, F.D., Venkatesh, V.: User acceptance of information technology: toward a unified view. MIS Q. **27**(3), 425–478 (2003)
9. Davis, F.D.: Perceived usefulness, perceived ease of use, and user acceptance. MIS Q. **13**(3), 319–339 (1989)
10. Davis, F.D., Bagozzi, R.P., Warshaw, P.R.: Extrinsic and intrinsic motivation to use computers in the workplace. J. Appl. Soc. Psychol. **22**(14), 1111–1132 (1992)
11. Ajzen, I., Ajzen, I.: The theory of planned behavior. Organ. Behav. Human Decis. Process. **50**(2), 179–211 (1991)
12. Davis, F.D., Bagozzi, R.P., Warshaw, P.R.: User acceptance of computer technology: a comparison of two theoretical models. Manag. Sci. **35**(8), 982 (1989)
13. Rogers, E.M.; Diffusion of Innovations (1995)
14. Bandura, A.: Social foundations of thought and action: a social cognitive theory (1986)
15. Thompson, R.L., Higgins, C.A., Howell, J.M.: Personal computing: toward a conceptual model of utilization. MIS Q. **15**(1), 125 (1991)

16. Chang, I.-C., Hwang, H.-G., Hung, W.-F., Li, Y.-C.: Physicians' acceptance of pharmacokinetics-based clinical decision support systems. Expert Syst. Appl. **33**(2), 296–303 (2007)
17. Neufeld, D.J., Dong, L., Higgins, C.: Charismatic leadership and user acceptance of information technology. Eur. J. Inf. Syst. **16**(4), 494–510 (2007)
18. Venkatesh, V., Thong, J., Xu, X.: Consumer acceptance and user of information technology: Extending the unified theory of acceptance and use of technology. MIS Q. **36**
19. Carlson, J.R., Robert W.: Channel expansion theory and the experiential nature of media richness perceptions. Zmud Sour. Acad. Manag. J. **42**(2), 153–170 (1999)
20. Mackay, W.: Diversity in the use of electronic mail: a preliminary inquiry. ACM Trans. Inf. Syst. **6**(4), 380–397 (1988)
21. Whittaker, S., Sidner, C.: Email overload: exploring personal information management of email. Proc. SIGCHI Conf. Hum. Factors Comput. Syst. Common Gr. **35**, 276–283 (1996)
22. Vishwanath, A., Herath, T., Chen, R., Wang, J., Rao, H.R.: Why do people get phished? Testing individual differences in phishing vulnerability within an integrated, information processing model. Decis. Support Syst. **51**(3), 576–586 (2011)
23. Shajari, M., Ismail, Z.: Jurnal Teknologi Constructing an adoption model for e-government services. J. Teknol. **2**, 29–37 (2014)
24. Grabner-Kräuter, S., Kaluscha, E., M. Fladnitzer, Perspectives of online trust and similar constructs: a conceptual clarification. In: 8th International Conference on Electronics and Commerce, pp. 235–243, Oct 2006
25. Pavlou, PA.: Consumer acceptance of electronic commerce: integrating trust and risk with the technology acceptance model. Int. J. Electron. Commer. **7**(3), 69–103 (2003)
26. D. H. McKnight, L. L. Cummings, and N. L. Chervany, "Initial Trust Formation in New
27. Carter, L., Bélanger, F.: The utilization of e-government services: citizen trust, innovation and acceptance factors. Inf. Syst. J. **15**(1), 5–25 (2005)
28. Costa, P.T., McCrae, R.R.: Professional manual: revised NEO personality inventory (NEO-PI-R) and NEO five-factor inventory (NEO-FFI). Odessa FL Psychol. Assess. Resour. **3**, 101 (1992)
29. McCrae, R.R., Costa, P.T.J.: Personality trait structure as a human universal. Am. Psychol. **52**(5), 509–516 (1997)
30. Hair, J.F., Blackk, W.C., Babin, B.J., Anderson, R.E.: Multivariate data analysis (6th ed.) Analysis, pp. 4–4 (2006)
31. Hair, J.F., Black, W.C., Babin, B.J., Anderson, R.E., Tatham R.L.: Multivariate Data Analysis (2010)
32. Nunnally, J., Bernstein, I.: Psychometric theory, 3rd edn., vol. 3, p. 701 McGraw-Hill, New York (1994)
33. Alseadoon, I., Chan, T., Foo, E., Nieto, J.G.: Who is more susceptible to phishing emails? A Saudi Arabian study. In: 23rd Australas. Conference on Information Systems Trusteer 2009, pp. 1–11 (2012)
34. Landers, R.N., Lounsbury, J.W.: An investigation of big five and narrow personality traits in relation to Internet usage. Comput. Human Behav. **22**(2), 283–293 (2006)

Perseverance of Observability in Hovering Quadcopter Under Sensor Faults

Mohammad Khoiron, Abdullah Alkaff, and A. K. Rusdhianto Effendie

Abstract Quadcopter is modeled as a cascade system by considering the rotor dynamics. The cascade systems of rotor plant and quadcopter plant are combined to form an augmented model of quadcopter. In hovering condition, the augmented model is represented as a linear model by taking linearization around operating states. The model is used to analyze the quadcopter observability and controllability. It is shown that the cascade systems are always controllable irrespective of the sensor faults. Perseverance of cascade system observability under certain sensor faults is shown. The rotor state observability under severe sensor faults is proved theoretically. Simulation is conducted to show the applicability of the method derived.

Keywords Quadcopter · Sensor failure · Cascade system · Controllability · Observability

1 Introduction

Quadcopter is an autonomous system that relies heavily on the health of its components, including sensors. The sensors in the quadcopter function as state observers knowing the external environment or the internal system states [1]. Quadcopter is often equipped by several sensors, on each of the rotor and the body of the quadcopter. Because the quadcopter has four rotors, then it requires four sensors, one for each rotor, while the main body of quadcopter itself is generally equipped with one IMU and one GPS receiver [2].

Issues that attracted researchers to study imperfect quadcopter usually are when the rotors are not perfect or experiencing faults. Freddi et al. [3] have presented that a quadcopter is not fully controllable although only one rotor is fully/partially failure. Du et al. [4] studied fault-tolerant control of quadcopter when experiencing multirotor failure. Saied et al. [5] studied nonlinear controllability in octocopter and

M. Khoiron (✉) · A. Alkaff · A. K. Rusdhianto Effendie
Institut Teknologi Sepuluh Nopember, Surabaya, Indonesia
e-mail: moh.khoironbenraxid@gmail.com

E. Joelianto et al. (eds.), *Cyber Physical, Computer and Automation System*, Advances in Intelligent Systems and Computing 1291,
https://doi.org/10.1007/978-981-33-4062-6_4

showed that the vehicle remains fully controllable although some of its rotor fails, where in [6] a fault diagnosis is also considered as main problem in fault-tolerant control. Lanzon et al. [1] and Mueller and Mueller et al. [7] have proposed strategy controls to compensate the rotor faults by giving up controlling one or more state of the quadcopter.

Theoretical study of observability preservation on dynamical system under sensor failure is thoroughly given by Commault et al. [8], and the sensor is classified as critical sensor and non-critical, where the critical sensor faults lead to loss of observability of the given dynamic system. Here, we adapt that research in our way on theoretical study of observability when quadcopter is in imperfect information, that is, when some of its sensors (rotor, GPS, IMU) are failed. Considering its flight environmental is uncertain [9], this issue becomes important and interesting. Theoretical study of observability when quadcopter is in imperfect information and therefore is a necessity. This issue becomes increasingly challenging when each of the rotor dynamics of the quadcopter is also considered, unlike the general assumption that rotor dynamics can be ignored.

The aims of this study are analyzing controllability and observability of the quadcopter experiencing sensor failure, by considering the quadcopter rotor dynamics. Together with quadcopter body dynamics, they form a cascade model for quadcopter dynamics which then can be combined to create an augmented model for the quadcopter. This augmented model will be used as the basis for the analysis. It will be shown that sensor faults do not inflict any change to the controllability, hence the partially observable system due to the sensor failure is still controllable. Perseverance of the rotor state observability is also proved under severe sensor faults.

2 The Dynamics of Quadcopter (QC) as Cascade System

2.1 *Models of QC Elements*

Including QC rotor, dynamics makes the quadcopter model becomes a cascade system where QC rotors input $u$is QC input, u QC rotor outpu, t $y_r$is QC body input, u_b, and QC body output is y_b QC output, as $u$given in Figure 1.

The four QC rotors can be represented in a state equation as follows. Define the QC rotor state as $x_r = (w_1, w_2, w_3, w_4) \in X_1$, dim $X_1 = 4$ and the QC rotor's input as $u_r = v_i, i = 1, 2, 3, 4$. Therefore, the state equation is given by

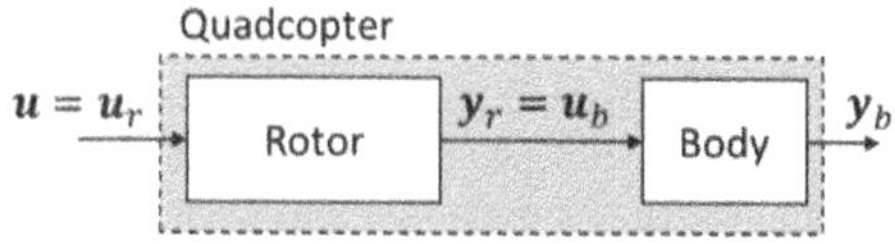

Fig. 1 Quadcopter as cascade system

$$
\begin{aligned}
x_r &= A_r X_r + B_r u_r \\
y_r &= 0f
\end{aligned} \tag{1}
$$

$$
= \begin{bmatrix} 1 & 1 & 1 & 1 \\ 0 & -1 & 0 & 1 \\ -1 & 0 & 1 & 0 \\ d & -d & d & -d \end{bmatrix} = \begin{bmatrix} f_1 \\ f_2 \\ f_3 \\ f_4 \end{bmatrix} = \begin{bmatrix} K_d\left(\omega_1^2 + \omega_2^2 + \omega_1^2 + \omega_4^2\right) \\ lK_d\left(-\omega_2^2 + \omega_4^2\right) \\ lK_d\left(-\omega_1^2 + \omega_3^2\right) \\ dK_d\left(\omega_1^2 - \omega_2^2 + \omega_1^2 - \omega_4^2\right) \end{bmatrix} \tag{2}
$$

where $A_r = \text{diag}\ (\tau_1, \ldots, \tau_2), B_r = \text{diag}\ (K_{r1}/\tau_1, \ldots K_{r4}/\tau_4) \cdot K_{ri}$ is QC rotor constant and is QC rotor torque, respectively, $f$is the produced forces, l distant from rotor 1 rotor 3 (or 2–4) and $d(= K_d/K_r)$ is drag constant.

State space representation of the QC body dynamics is given as follows. Define $x_b = \left(z, \dot{z}, \varphi, \dot{\varphi}, \theta, \dot{\theta}, \psi, \dot{\psi}\right) \in X_2$, $\dim X_2 = 8$ as the QC body state and $u_b = (u_1, u_2, u_3, u_4)$ as its input, therefore the QC body's state equation is given by

$$
\begin{aligned}
x_b &= f_b(x_b) + g_b(x_b)u_b \\
y_b &= H_b x_b
\end{aligned} \tag{3}
$$

where

$$
f_b(x_b) = \begin{bmatrix} \dot{z} \\ -(K_{tz}\dot{z}/m) - g \\ \dot{\theta}\dot{\psi}\left(1_y - 1_z\right)/1_x - K_{rx}\ddot{\varphi}^2 \\ \dot{\theta}\dot{\psi}(1_z - 1_x)/1_y - K_{ry}\ddot{\theta}^2 \\ \dot{\varphi}\dot{\theta}\left(1_x - 1_y\right)/1_z - K_{rz}\ddot{\psi}^2 \end{bmatrix} \tag{4}
$$

$$
g_b(x_b) = \begin{bmatrix} 0 & 0 & 0 & 0 \\ (\cos\varphi\cos\theta)/m & 0 & 0 & 0 \\ 0 & 0 & 0 & 0 \\ 0 & 1/l_x & 0 & 0 \\ 0 & 0 & 0 & 0 \\ 0 & 0 & 1/1_y & 0 \\ 0 & 0 & 0 & 0 \\ 0 & 0 & 0 & 1/1_z \end{bmatrix} \tag{5}
$$

Y_b is an eight-dimensional QC body output vector and $H_b = I_8$.

2.2 Augmented Model of Quadcopter

The QC rotor and the QC body can be represented in an augmented state equation as follows. Define the augmented state as $x - \begin{bmatrix} x_r \\ x_b \end{bmatrix} \in X_1 \oplus X_2 = X$. The relationship between the QC rotor and body is already mentioned above that is $u = u_r, y_r = u_b$dan $y = y_b$, $\dim u = 4$and $\dim y = 8$. The augmented state equation is given by

$$\begin{aligned} \dot{x} &= f(x) + gu \\ y &= Hx \end{aligned} \tag{6}$$

where

$$f(x) - \begin{bmatrix} A_r \\ g_b(x_b)\phi(x_r) + f_b(x_b) \end{bmatrix} \tag{7}$$

$$g - \begin{bmatrix} B_r \\ 0 \end{bmatrix} \tag{8}$$

$$H = \begin{bmatrix} 0 \ H_b \end{bmatrix} \tag{9}$$

$B_r$and H_b are given in (1) and (3), respectively.

2.3 Linear Model for Hovering Quadcopter

The analysis is carried out for hovering quadcopter that is when $\varphi = \dot{\varphi} = \theta = \dot{\theta} = \psi = \dot{\psi} \approx 0°, \dot{z} \approx 0$and $u_{b1} = mg\left(\omega_1 = \sqrt{mg/4K_p}\right)$ [10], and the change in QC body states is small. Let us take $\varphi = \dot{\varphi} = \theta = \dot{\theta} = \psi = \dot{\psi} \approx 0°, \dot{z} \approx 0$ as $x°$ and $\omega_i = \sqrt{mg/4K_p}$ as ω^*. Taylor expansion is used for linearization, and the resulted linear model of the augmented state equation is

$$\begin{aligned} \dot{x} &= A_x + B_u \\ y &= H_x \end{aligned} \tag{10}$$

where

$$A = \mathrm{d}(f(x))/\mathrm{d}x|_{x^*,\omega^*} = \begin{bmatrix} A_r & 0 \\ A_{rb} & A_b \end{bmatrix} \tag{11}$$

$$B = \mathrm{d}(g)/\mathrm{d}u|_{x^*,\omega^*} = g \tag{12}$$

A_r and A_b are given in (1) and

$$A_b = \mathrm{d}(f_b(x_b))/\mathrm{d}u_b|_{x^*} \tag{13}$$

$$A_{rb} = \mathrm{d}(g_b(x_b))/\mathrm{d}u_b|_{x^*}\mathrm{d}(\phi(x_r))/\mathrm{d}x_r|_{\omega^*} = B_b H_r \tag{14}$$

H is already linear as given in (9).

3 Effect of Sensor Faults

Sensor faults directly affect the observability of the quadcopter. In other word, when the quadcopter is not experiencing sensor faults, the quadcopter can be said to be observable. However, in this study, there is no QC rotor sensor, or all the QC rotor sensors fail. Therefore, all states of the QC rotor must be obtainable from the available IMU sensor. As for the IMU faults, there are two possibilities that we propose that are as follows:

- The accelero sensor is failure but the gyro sensor is fine.
- The gyro sensor is failure but the accelero sensor is fine.

Two difficulties arise, (1) the unavailability of some of the states of the QC body, and (2) the calculation of the QC rotor states which is highly dependent on the availability of the QC body states. Hence, it is necessary to understand the indirect relationship between the QC rotor states and the QC body states which is given in [11].

$$u_{b,i}(t) = -B_b^{'}\mathrm{e}^{A_b(t1-t)}W_c^{-1}(t_1)\left[\mathrm{e}^{A_b(t_1)}x_{0,b} - x_{1,b}\right], i = 1, 2, 3, 4 \tag{15}$$

This equation is used to calculate an appropriate input of the QC body given the corresponding output/states of the. This calculation can be performed under a condition that the QC body is controllable and the information $\left(x_{0,b}, x_{1,b}\right)$ is available. The QC rotor states then can be obtained by solving (2). But with the failure of some part of the IMU sensor, the information required is partially available and, therefore, not all $u_b$can be calculated which in turn making matrix transformation in (2) is no longer non-singular. Coupled with the needs of a unique mapping in observability, modification is needed for (2) so it can be solved using pseudoinverse [12].

4 The Controllability Analysis of Cascade System in Hovering Quadcopter

The analysis of controllability of cascade system is carried out by analyzing controllability of each subsystem. The cascade system is controllable if each subsystem is controllable, and there exists an output feedback law so the resulting system is controllable as given in [13] and [14].

Lemma 1 *If the QC body satisfies controllability rank condition (C.R.C.) at* $\left(\varphi = \dot{\varphi} = \theta = \dot{\theta} = \psi = \dot{\psi} \approx 0^\circ\right)$ *and the QC rotor is controllable, then there exist* $\phi : x_r \to u_b$ *so that the quadcopter as given in* (10) *satisfies C.R.C.*

Proof The QC body in hover condition is given by

$$\begin{aligned} \dot{x}_b &= A_b x_b + B_b u_b \\ y_b &= H_b x_b \end{aligned} \tag{16}$$

where A_b B_b and H_b are given by (11), (14) and (3), respectively. The QC body is controllable everywhere in because for the QC body controllability matrix,

$$C_b = \left[\, B_b \; A_b B_b \; A_b^2 B_b \; A_b^3 B_b \; \cdots \; A_b^7 B_b \,\right] \tag{17}$$

has $\rho(C_b) = 8$. Similarly, for the QC rotor is given in (1). The QC rotor is controllable everywhere in because the QC rotor controllability matrix

$$C_r = \left[\, B_r \; A_r B_r \; A_r^2 B_r \; A_r^3 B_r \,\right] \tag{18}$$

has $\rho(C_r) = 4. \phi : x_r \to u_b$ as is given in (2). Hence, by Theorem 1, the resulting system as given in (10) is controllable.

5 The Observability Analysis of Cascade System in Hovering Quadcopter

In this study, we assume that the loss observability only occurs when some of the sensor fails [8]. Because there is no sensor QC rotor, perseverance of the observability then might be considered as the ability to calculate the QC rotor states given the available QC body states from its functioning, denoted by $\varphi : y_b \to x_r$, φ is injective, which can be obtained by utilizing (15) and (2). From these arguments, we can characterize the observability analysis in the studied system as follows:

Proposition 1 *In the absence of the QC rotor sensor, the observability of the quadcopter can be preserved, if and only if the QCs body is controllable and observable,* *and there is inverse function* $\varphi : y_b \rightarrow x_r \in R^4$, φ *is injective so that the quadcopter as given in* (10) *is observable.*

5.1 The Case of Perfect IMU Sensor

If the IMU sensor is perfect, the ouput matrix of the QC body is the same as given (10), so all the QC body states, $(z, \dot{z}, \varphi, \dot{\varphi}, \theta, \dot{\theta}, \psi, \dot{\psi})$ can be obtained by the sensors.

Lemma 2 If IMU is perfect, then the quadcopter is observable.

Proof The QC body is observable because its observability matrix

$$O_b = \left[H_b \; H_b A_b \; H_b A_b^2 \; H_b A_b^3 \cdots H_b A_b^7 \right]' \tag{19}$$

has $\rho(O_b) = 8$ the QC rotor is observable because from the available the QC body state, there exists $\varphi : y_b \rightarrow X_r$, which is given by (15) and the inverse of (2). Hence, Lemma 2 is proofed.

5.2 The Case of Failure in Accelero Sensor

When the gyro sensor is the only functioning sensor, the available QC body states are $(\varphi, \dot{\varphi}, \theta, \dot{\theta}, \psi, \dot{\psi})$. However, hovering condition requires $b_1 = mg$ and is steady in a certain value. Therefore, z can be obtained using z dynamics of the QC body as given by

$$\ddot{z} = (\sin\varphi \cos\theta) u_{b_1} / m \tag{20}$$

Lemma 3 Quadcopter with only gyro sensor is observable.

Proof Given $u_{b_1} = mg$ and (20), the available QC states are as in the case of perfect IMU. Therefore, the quadcopter is observable.

5.3 The Case of Failure in Gyro Sensor

When the accelero sensor is available, in addition to $(z, \dot{z})$, we can also obtain $(\varphi, \dot{\varphi}, \theta, \dot{\theta})$ as given by [15]

$$\tan\varphi = [\ddot{y}/\ddot{z}] \tag{21}$$

$$\tan\theta = [-\ddot{x}/\ddot{y}\sin\varphi + \ddot{z}\cos\varphi] = -\ddot{x}/\sqrt{\ddot{y}^2 + \ddot{z}^2} \tag{22}$$

However, it is impossible to obtain $(\psi, \dot{\psi})$. The loss information of $(\psi, \dot{\psi})$ causes the QC body unobservable. Hence, the quadcopter is unobservable. Still, the QC rotor states can be calculated from $(z, \dot{z}, \varphi, \dot{\varphi}, \theta, \dot{\theta})$ as follows. Using (15), the corresponding $(u_{b_1}, u_{b_2}, u_{b_3})$ from$(z, \dot{z}, \varphi, \dot{\varphi}, \theta, \dot{\theta})$ can be obtained. Then, the force generated by the QC rotor (or the QC rotor state itself) can be calculated by solving (2) using pseudoinverse. Although the observability of the quadcopter cannot be preserved because the QC body is unobservable, the observability of the QC rotor can be maintained. The following lemma is therefore proved.

Lemma 4 The observability of the QC rotor is preserved with only gyro sensor.

6 Simulation

To validate the theoretical results, simulations based on all IMU sensor conditions have been conducted. Parameters for simulation are taken from [13]. The first simulation is for perfect IMU or IMU with accelero sensor fails. The set point for is z $10m$ and the Euler angles, (φ, θ, ψ), is set to follow a given pattern. The results are presented in Fig. 2a. It can be seen that all angular Euler angles can follow the set point and steady in a relatively fast time frame. The second simulation is when the gyro sensor fails. The results are given in Fig. 2b. Because $\psi, \dot{\psi}$ is not available, we omit the need to control yaw angle and the set points for (φ, θ) are same with the first simulation. Although the information is incomplete, the quadcopter still can follow the given set points. Figure 3 compares the results of the calculation of the QC rotor state with its real value. It shows that the two calculations are same with those original value.

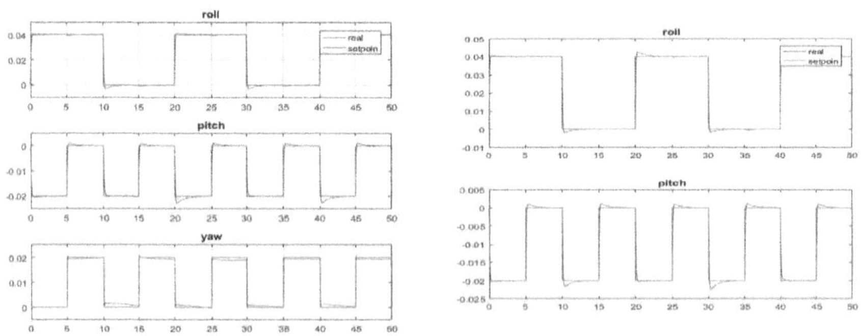

(a) (b)

Fig. 2 **a** When IMU is perfect or under accelero sensor failure, **b** under gyro sensor failure

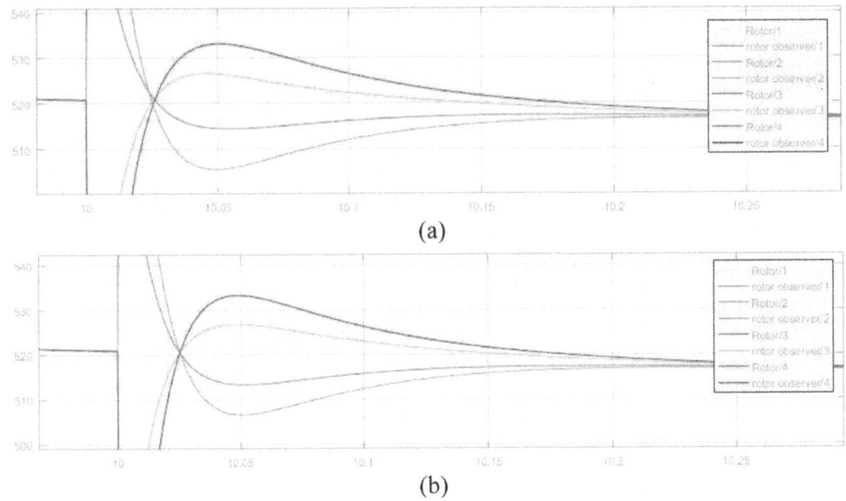

Fig. 3 Comparison of the real value and calculation of the QC rotor states, **a** when IMU is perfect or under accelero sensor failure, **b** under gyro sensor failure

7 Conclusion

We have presented the controllability and observability of quadcopter under sensor faults with the inclusion of rotor dynamics. First, dynamics of quadcopter cascade system have been derived. Second, variations of IMU faults have been considered. Controllability of the system is not affected by the sensor faults, while observability is strongly influenced by it. While the rotor plant is always observable under various sensor fault conditions considered, observability of the system can be preserved under accelero sensor faults and absence of rotor sensor. The simulation shows the accuracy of the proposed method. The future work will focus on extending the analysis to the more general quadcopter maneuvers.

References

1. Freddi, A., Lanzon, A., Longhi, S.: A feedback linearization approach to fault tolerance in quadcopter Vehicles. The IFAC (2011)
2. Bresciani, T.: Modelling identification and control of a quadcopter helicopter. LUP Student Papers (2008)
3. Freddi, A., Lanzon, A., Longhi, S.: Flight control of quadcopter vehicle subsequent to a rotor failure. J. Guid. Contr. Dyn. 580–591 (2012)
4. Du, G.-X., Quan, Q., Cai, K.-Y.: Controllability analysis and degraded control for a class of hexacopters subject to rotor failures. J. Intell. Robot. Syst. 143–157 (2015)
5. Saied, M., Lussier, B., Fantoni, I., Shraim, H., Francis, C.: Local controllability and attitude stabilization of multirotor UAVs: validation on a coaxial octorotor. Robot. Auton. Syst. (2017)

6. Saied, M., Lussier, B., Fantoni, I., Shraim, H., Francis, C: Fault diagnosis and fault-tolerant control of an octorotor UAV using motors speeds measurement. IFAC papersOnLine, pp 5263–5268 (2017)
7. Mueller, M., D'Andrea, R.: Stability and control of a quadcopter despite the complete loss of one, two, or three propeller. IEEE International Conference on Robotics and Automation (2014)
8. Commault, C., Dion, J.-M., Trinh, D.H.: Observability preservation under sensor failure. IEEE Trans. Autom. Contr. pp 1554–1559 (2008)
9. Zohdi, T.I.: On the dynamics and breakup of quadcopters using a discrete element method framework. Comput. Methods Appl. Mech. Eng. 503–521 (2017)
10. Leong, B.T.M., Low, S.M., Ooi, M.P.-L.: Low-cost microcontroller-based hover control design of a quadcopter. Proced. Eng. **41**, 458–464 (2012)
11. Chen, C.-T.: Linear System Theory and Design. 1st edn. Oxford University Press, Inc (1999)
12. Klema, V.C., Laub, A.J.: The singular value decomposition and some applications. Proced. Eng. (1980)
13. Kalouptsidis, N., Tsinias, J.: Controllable cascade connections of nonlinear systems. Nonlin. Anal. Theory, Methods Appl. 1929–1244 (1987)
14. Tsinias, J., Kalouptisidis, N.: Output feedback design and controllable cascade connections of nonlinear system. Syst. Contr. Lett. **2**, 230–236 (1982)
15. Shuster, M.D., Markley, F.L.: General formula for extracting the euler angles. J. Guid. Contr. Dyn. (2006)
16. Zeghlache, S., Kara, K., Saiga, D.: Fault tolerant control based on interval type-2 fuzzy sliding mode controller for coaxial Trirotor aircraft. ISA Trans. 215–231(2015)

Integration of FIR and Butterworth Algorithm for Real-Time Extraction of Recorded ECG Signals

Mardi Turnip, Abdi Dharma, Andrian, Adam Afriansyah, Ade Oktarino, and Arjon Turnip

Abstract A heart attack is an emergency condition that needs to be treated as soon as possible. Limitations of tools, busyness, and distance from the hospital often make a patient unaware of the existence of heart disease. In this study, a real-time remote monitoring of cardiac activity is being developed. Integration of FIR and Butterworth algorithm for real-time ECG signals extractions is being proposed. Quality testing of the instrument was carried out on eight subjects in the conditions of rest, walking, and jogging. From the results of data processing, it was obtained that the proposed algorithm was able to extract features well with little delay.

Keywords ECG · Real time · Butterworth · Extraction · Heart attack

1 Introduction

Heart attack is one of the most deadly diseases and has the highest death toll of the victim [1–4]. Unhealthy and destructive lifestyles can make anyone from various backgrounds suffer a heart attack. This disease often has no symptoms that are invisible, so patients do not realize it. As we know, the heart is the most vital organ in humans because the body will not operate without it. Heart health conditions are not easy to recognize from the outside. In general, an unhealthy heart will not show symptoms before it reaches a severe condition. There are also certain heart conditions that are not normal, but not harmful to health. However, this does not mean that it

M. Turnip · A. Dharma · Andrian
Faculty of Technology and Computer Science, Universitas Prima Indonesia, Medan, Indonesia

A. Afriansyah · A. Oktarino
Faculty of Technology and Computer Science, Universitas Adiwangsa Jambi, Jambi, Indonesia

A. Turnip (✉)
Department of Electrical Engineering, Universitas Padjadjaran, Sumedang, Indonesia
e-mail: jujhin@gmail.com; turnip@unpad.ac.id

E. Joelianto et al. (eds.), *Cyber Physical, Computer and Automation System*, Advances in Intelligent Systems and Computing 1291,
https://doi.org/10.1007/978-981-33-4062-6_5

can be ignored, because some of these abnormal conditions require special attention to last longer.

Heart health is the main reason for people that must regularly control their body health condition to the doctor, not only for patient who have heart problems, but also for those who have the potential to contract heart disease. The problem is that not everyone can allocate their time and have enough financial support to routinely control to a doctor. Electrocardiogram (ECG) is one way to find out the condition of the heart [5–8]. The ECG signals indicate electrical activity that occurs in the human heart. The ECG signals can provide complete information to determine a person's heart condition. With the aim that ECG tools can be used by more people, the authors developed an ECG program to be applied to cheaper BITalino tools.

Various studies related to the development of cardiac monitoring devices with ECG have been widely developed [9–17]. However, the majority of these studies was carried out offline. Development of ECG in real time is still rarely done, especially the use of FIR and Butterworth filter algorithms. In this study, an application that can implement algorithms to detect ECG signals with a BITalino tool in real time was developed. The filter application is designed to be applied to the BITalino tool with the Python programming language. The proposed adaptive digital filtering algorithm enables real-time monitoring and adapting to various noise arising from the body's internal activities as well as the physical activity of the subject. As a step to overcome the problem of noise pollution on ECG recordings, the author will implement both the finite impulse response (FIR) algorithm and Butterworth filter to the application to process the ECG data. The application will read the ECG recording files that have been recorded by BITalino and will work to filter the data and produce CSV (coma separated value) type data. CSV file will contain the frequency value that has been processed and can be used for further processing or directly displayed in graphical form for visual purposes.

2 Methods

To determine the condition of the subject's heart health, things to note are the characteristics of the ECG signal recording or also called diagnostic features. A normal ECG consists of a P wave, a QRS complex, and a T wave as shown in Fig. 1. Characteristics of the ECG are normal if the signal behavior, such as the duration of time of each wave and between waves, amplitude, heart rhythm, and so on.

Various studies related to the development of cardiac monitoring devices with ECG have been widely developed [9–17]. However, the majority of these studies was carried out offline. Development of ECG in real time is still rarely done, especially the use of FIR and Butterworth filter algorithms. In this study, an application that can implement algorithms to detect ECG signals with a BITalino tool in real time was developed. The filter application is designed to be applied to the BITalino tool with the Python programming language. The proposed adaptive digital filtering algorithm enables real-time monitoring and adapting to various noise arising from the body's

Fig. 1 General form of ECG signals

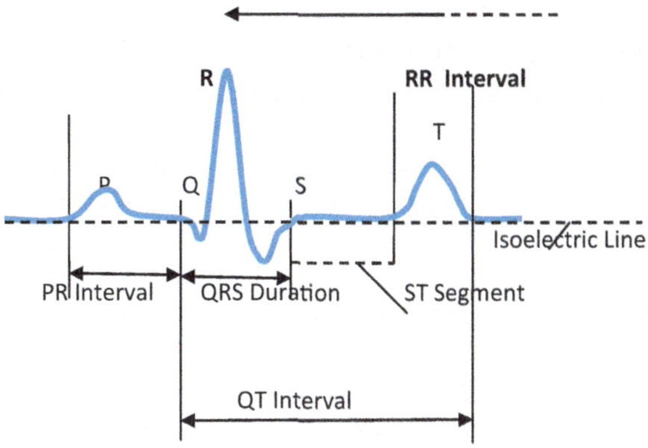

internal activities as well as the physical activity of the subject. As a step to overcome the problem of noise pollution on ECG recordings, the author will implement both the finite impulse response (FIR) algorithm and Butterworth filter to the application to process the ECG data. The application will read the ECG recording files that have been recorded by BITalino and will work to filter the data and produce CSV (coma separated value) type data. The CSV file will contain the frequency value that has been processed and can be used for further processing or directly displayed in graphical form for visual purposes.

$$\begin{aligned} y(n) &= b(1)x(n) + b(2)x(n-1) + L \\ &\quad + b(nb+1)x(n-nb)a(2)y(n-1) \\ &\quad - L - a(na+1)y(n-na) \end{aligned} \tag{1}$$

where $x(n)$ is the input, $y(n)$ is the output, constants $b(i)$ and $a(i)$ are the filter coefficients, then na and nb are the maximum filter orders. Various kinds of filter names are used to describe the filter depending on the number na and nb. If $nb = 0$, filter is often called infinite impulse response (IIR), all-pole, recursive or autoregressive (AR). If $na = 0$, filters are often called finite impulse response (FIR), all-zero, non-recursive or moving average (MA). If both are greater than zero (na and nb are greater than zero), filters are often called IIR, pole zero, recursive, or autoregressive moving average (ARMA) [18]. The finite impulse response (FIR) algorithm is a filter that has an impulse response to finite input. The basic characteristics of the FIR filter are shown in Eq. 2 [19, 20].

$$y(k) = \sum_{k=0}^{N-1} h(k)x(n-k) \tag{2}$$

where $h(k)$ is the frequency coefficient of the FIR filter response. This makes the FIR filter less sensitive or less influenced by the accuracy of the content compared to the similar IIR filter. In general, the range of N that is often used is 2^k. The value of the constant k is the minimum value that satisfies the rule of equation $N \leq 2^k$ becomes $k = [1 + \log\_2N]$. The Butterworth filter algorithm is a filter for processing signals with a more stable response frequency. The implementation of Butterworth filter is based on the bilinear transformation method. Butterworth filter can function as a stable amplifier for data retrieval in real time.

The ECG recording process was carried out with a special machine that can detect electrical impulses or often called electrocardiographs. The ECG recording was done within a certain time period in order to be able to get the subject's heart cycles needed for diagnosis. The process was done non-invasively, by attaching electrodes to the surface of the subject's chest, hands, or feet. One of the ECG recording devices that can be used is BITalino [21]. BITalino is a biomedical signal recording device that can be used anywhere because of its size which is easy to carry and its operation is quite practical. In general, the practice of physiological programming has special needs, such as equipment with certain specifications which usually have limited access or high prices. The BITalino was designed based on these problems so that biomedical signals can be more easily researched and developed.

The steps in implementing an algorithm in a preprocessing system of a signal recording to get signal data that is ready to be displayed or processed were the initial step in the system starts with retrieving ECG recording data within a certain time with the BITalino tool for processing; data was converted into arrays so that it can be processed with a filter algorithm; for the filter process, some coefficients were determined based on previous research recommendations, and the rest depends on the desired conditions; Based on the functions and characteristics of the chosen algorithm, the process begins by first applying the FIR filter. The filter process will produce signal data in the form of arrays; the array data derived from the FIR filter will be continued along with the parameters used in the Butterworth filter; the results obtained in the form of data arrays in the form of signal frequencies that can already be displayed in graphical form for visual purposes, or proceed to the next stage in the form of arrays. The recording was done with a number of conditions, one of which is used was a rest condition that was sitting position and not doing other physical activities. In general, this condition was used as a parameter indicating the normal state of a person's heart. With the Python programming language, the proposed system was designed supported by several complementary modules from the library used, such as the BITalino connection module, Matplotlib for supplementary graphics, and SciPy for signal processing.

3 Results and Discussions

Program simulation results using subject ECG records under rest for 5 s from 2 min is shown in Fig. 2. The frequency of the sample used in the experiment was 1000 Hz, as a result the total amount of data displayed in 2 min becomes too dense to be evaluated. So as a comparison of results, it was displayed for the first 5 s where two data conditions are shown, namely the first condition as the initial condition before

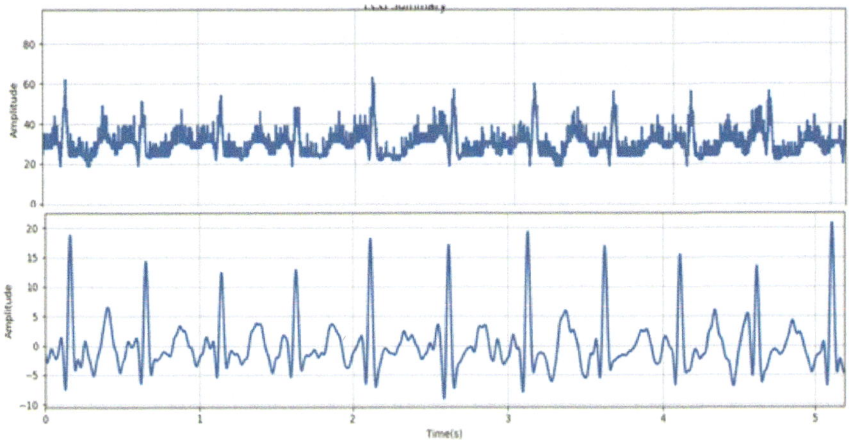

Fig. 2 Raw data and filtered signals results

the signal data was processed (raw data) and followed by the condition when the signal has been processed (filtered) with the FIR algorithm and Butterworth filter. Figure 2 shows a display graph of signal waves in the *x*- and *y*-axes. The *y*-axis is the signal amplitude, where the initial shift occurs so it is not based on 0. This shows that the signal has been polluted by noise, one of which is the baseline wander. In addition, the magnitude of the amplitude was also not reasonable with a range of 20–60 dB, which is not a normal size ECG. The ECG waves also found negative peaks to allow amplitudes of values lower than 0, but in the raw condition, all signal wave positions were above of 0. Negative values on the ECG were also parameters of a person's heart condition that cannot be ignored.

The signal was processed first by FIR and then followed by Butterworth which has two gradual processes, namely low-pass and high-pass filters. Each process was more focused in eliminating certain types of noise, which is also the reason why it is necessary to combine more than one algorithm. Thus, signal data that can be processed also has a wider range. As a comparison, the signal processing process is given in Fig. 3 for each algorithm process. In Fig. 3, there are two signal conditions with the duration of the first 5 s, the signal source, the condition, the taken axis that same as Fig. 2 so that it can be used as a comparison. The first condition indicates that the signal was only processed by the FIR algorithm. The data shows a significant change from raw conditions that have many artifacts and other disturbances. But the signal position is still in the range of 15–40 dB which is not normal. In addition, the signal data still has a lot of small wave indentations as a sign that there is still noise content. In the second condition, in Fig. 3, the signal was only processed with the Butterworth algorithm for the first 5 s. The display shows that the signal is in a better position, with fairly stable and smooth wave characteristics. But if we pay attention, some wave peaks become indecisive which will be a barrier when we want to extract

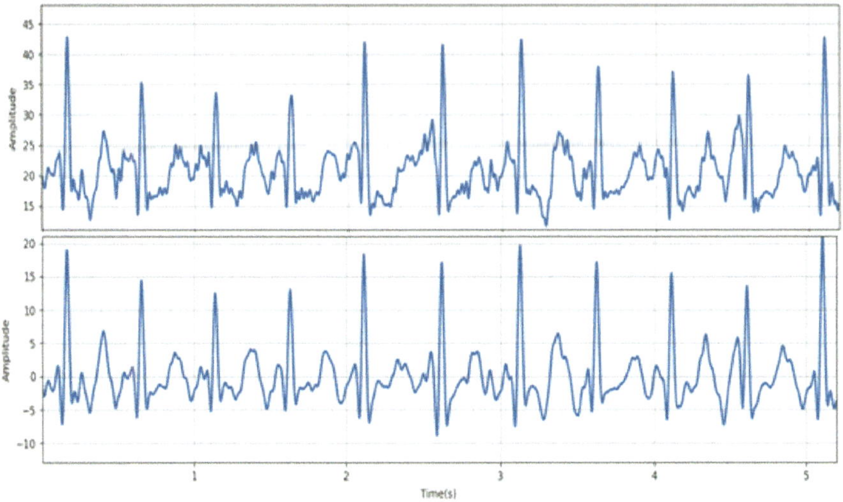

Fig. 3 Results of FIR and Butterworth method, respectively

the ECG feature. This is another reason to combine the two algorithms, because each algorithm has focus and shortcomings.

The parameters used in the filter design both FIR and Butterworth were recommendations, both from the results of the study of the parameters and recommendations for signal compatibility with recording devices, namely BITalino. Basically, digital filters have a way of working that is almost the same to each other only differentiated by design and parameters. For example, the sampling rate is the frequency of the sample to be chosen. BITalino allows 1, 10, and 1000 Hz. The signal recording dataset used was a CSV file that records ECG electrical impulses. Because the recorded data has a large quantity, the signal data used in research is often the result of generation (artificial). One of the modules in Python that allows this is the SciPy module. BITalino devices record signals that are not just ECGs. Each time recording, all ports will work to record the impulse that was also read by electrodes, both ECG, EEG, and EMG. The ECG data is in the last column of the input array.

For FIR and Butterworth algorithm applications, programming was conducted in Python with the execution stages: import module/library; take the dataset; signal graphic; and filter section. The first thing to do when starting the system is to insert or implement modules and libraries in Python. In addition to basic functions, there are some functions that are also needed from other research fields, for example NumPy (mathematical), or Matplotlib (graph). Signal data to be processed as input needs to be determined in advance. In general applications, signal data has several file types depending on the recording media and the method to be selected for processing the data, for example .json, .txt, .csv, and so on. In this float, the system reads the coma separated value (csv) recording file which is also recorded with the BITalino module using the Python programming language. Besides functioning to eliminate noise in the signal data, which will be needed in processing ECG records, the system also allows the appearance of processed data in graphical form. This will facilitate the assessment of the accuracy and success rate of the algorithm. Figure 4 is a graph display of the results of the built-in filter algorithm. Furthermore, the filter function was neatly designed and arranged so that it can be called easily for various purposes such as modification. Figure 5 is the appearance of the filter section.

Here is how to execute commands and run filter processes on the system using the built-in FIR and Butterworth algorithms. The filtering process was done by reading the recorded CSV file dataset BITalino tool by entering the command in the console.

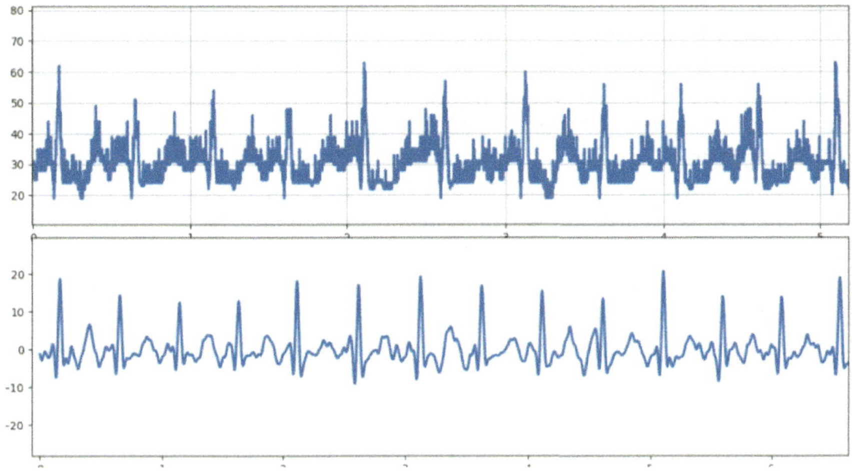

Fig. 4 Processed signal graph

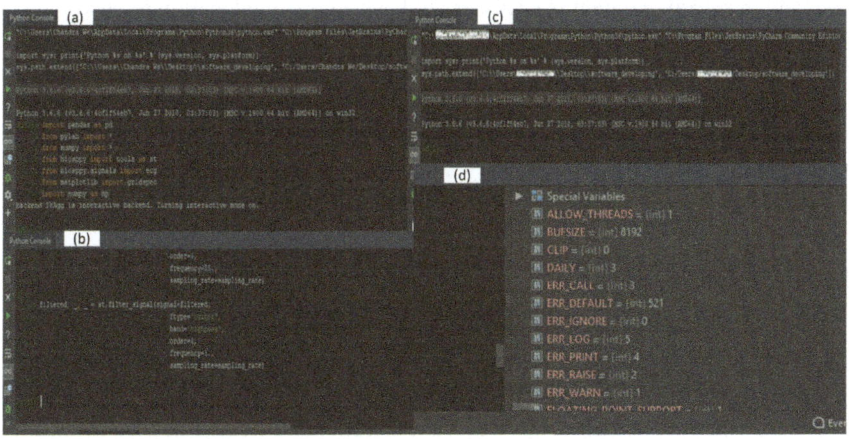

Fig. 5 **a** Filter section script, **b** display successful execution, **c** python console, and **d** display of variables

This section is provided to execute existing commands without the need for the design of the whole application and has a role as a place for input and output commands which are a form of interaction between the user and the system. This will facilitate and speed up the implementation process to be more effective. Because in addition to being a place of execution, the results and errors obtained are also displayed so that the user can immediately know. After executing the command code into the console, a new line will appear for the new command as shown in Fig. 5b. This shows that the command was successfully executed. In addition, other alternatives can also be used by adding a pointer to the code that will indicate the program running process. To see whether the process runs as desired and gets the intended results, checking variables can be done. Variables consist of coefficient variables and input parameter values, and outcome value variables from the process are carried out. All values have been stored in the variables section as shown in Fig. 5d which can be displayed either in an array, ArrayList, coefficient, and type of data. After executing the command and retrieving the data needed for processing and other things, the system also provides a file in the form of .csv. The file storage location will and must be the same as the program file location.

Basically, the system design was done mainly to implement the filter problem with the FIR and Butterworth algorithm. The ECG signal processing process itself still has many stages needed to be ready for use. The author successfully implemented the algorithm chosen for the original recording file that came from the BITalino tool with three types of activities, namely sitting as a rest condition, running normally, and walking fast with a little running. By focusing on the preprocessing and filtering stages, the disadvantages of implementing the system with the FIR and Butterworth algorithm in time are as follows: When the graph is displayed, the image will be static. Thus, if the data is large, the graph will look very dense and its characteristics cannot be considered. Ideally, the graph can move with a window that simply displays detailed ECG wave characteristics; the file that is read is still limited to CSV extension, and in selecting data it is necessary to have a clear file location to be known in order execution; there is still a delay of several seconds (delay) in the process because the system works by processing the entire data first before displaying it. This will be seen clearly when the data to be processed is large. Because the purpose of system implementation is for monitoring, it is expected that the process can be real time or have the smallest possible delay.

4 Conclusion

The FIR and Butterworth filter algorithms are built and can be applied simultaneously to complement each other on ECG recording data. Preprocessing process that does not run well will cause data shifting so it becomes inaccurate. The obtained filter results from the proposed algorithm at the preprocessing stage are stable with visible and distinguishable ECG wave features.

Acknowledgements This research was supported by Technical Implementation Unit for Instrumentation Development, Indonesian Institute of Sciences, funded by RISTEKDIKTI by INSINAS 2019, Indonesia.

References

1. Satija, U., Ramkumar, B., Manikandan, M.S.: Automated ecg noise detection and classification system for unsupervised healthcare monitoring. IEEE J. Biomed. Health Inf. **22**(3), 722–732 (2018)
2. Moeyersons, J., et al.: Artefact detection and quality assessment of ambulatory ECG signals. Comput. Methods Programs Biomed. **182** (2019)
3. Rachim, V.P., Chung, W.Y.: Wearable noncontact armband for mobile ECG monitoring system. IEEE Trans. Biomed. Circ. Syst. **10** (2016)
4. Majumder, S., Chen, L., Marinov, O., Chen, C.H., Mondal, T., Deen, M.J.: Noncontact wearable wireless ECG systems for long-term monitoring. IEEE Rev. Biomed. Eng. **11**(10), 306–321 (2018)
5. Bhaskar, P.C., Uplane, M.J.: High frequency electromyogram noise removal from electrocardiogram using FIR low pass filter based on FPGA. Proced. Tech. **25**, 497–504 (2016)
6. Choi, Y.J., Lee, J.Y., Kong, S.H.: ECG driver measurement system with conductive fabric-based dry electrodes. IEEE Access **6** (2017)
7. Kraus, M.S., Rishniwc, M., Diversc, T.J., Reefb, V.B., Gelzera, A.R.: Utility and accuracy of a smartphone-based electrocardiogram device as compared to a standard base-apex electrocardiogram in the horse. Veter. Sci. **125**, 141–147 (2019)
8. Safri, Nishfa Dewi, W., Erwin, : Analysis of electrocardiogram recording lead ii in patients with cardiovascular disease. Enfermería Clínica **29**(1), 23–25 (2019)
9. Fensli, R., Gundersen, T., Snaprud, T., Hejlesen, O.: Clinical evaluation of a wireless ECG sensor system for arrhythmia diagnostic purposes. Med. Eng. Phys. **35**(6), 697–703 (2013)
10. Søgaard, P., et al.: Transmission and loss of ECG snapshots: remote monitoring in implantable cardiac monitors. J. Electrocardiol. **56**, 24–28 (2019)
11. Faganeli Pucer, J., Kukar, M.: A topological approach to delineation and arrhythmic beats detection in unprocessed long-term ECG signals. Comput. Methods Programs Biomed. **164**, 159–168 (2018)
12. Kaminski, M., Prymas, P., Konobrodzka, A., Filberek, P., Sibrecht, G., Sierocki, W., Osinska, W., Wykretowicz, A., Lobodzinski, S., Guzik, P.: Clinical stage of acquired immunodeficiency syndrome in HIV-positive patients impacts the quality of the touch ECG recordings. J. Electrocardiol. **55**, 87–90 (2019)
13. Sharma, M., San Tan, R., Rajendra Acharya, U.: Automated heartbeat classification and detection of arrhythmia using optimal orthogonal wavelet filters. Informat. Med. Unlocked **16** (2019)
14. Tripathy, R.K., Paternina, M.R.A., Arrieta, J.G., Zamora-Méndez, A., Naika, G.R.: Automated detection of congestive heart failure from electrocardiogram signal using Stockwell transform and hybrid classification scheme. Comput. Methods Programs Biomed. **173**, 53–65 (2019)
15. Haotian, S., Haoren, S., Yixiang, H., Liqun, Z., Chengjin, Q., Chengliang, L.: A hierarchical method based on weighted extreme gradient boosting in ECG heartbeat classification. Comput. Methods Programs Biomed. **171**, 1–10 (2019)
16. Kruger, G.H., Latchamsetty, R., Langhals, N.B., Yokokawa, M., Chugh, A., Morady, F., Oral, H., Berenfeld, O.: Bimodal classification algorithm for atrial fibrillation detection from m-health ECG recordings. Comput. Biol. Med. **104**, 310–318 (2019)

17. Little, N.J., Shure.: MathWorks, Signal processing toolbox for use with matlab. User's Guide The Mathworks, pp 13. Inc. Loren (1993)
18. Kunaryo, H.B.: Aplikasi Tapis Adaptif FIR Untuk Menghilangkan Artefak Pada Sinyal Elektrokardiograf. http://eprints.undip.ac.id/25290/1/ML2F302468.pdf. Last accessed 2019/5/10
19. Ifeachor, C.E., Jervis, B.W.: Digital signal processing a practical approach Wokingham. Addison-Wesley Publishing Company (1993)
20. Butterworth, S.: On the theory of filter amplifiers. Wirel. Eng. **7**, 536 (1930)
21. Da Silva, H.P.: Biosignals for everyone. IEEE Pervas. Comput. **13**(4), 64

Multi-network Transmission Using Socket Programming to Support Command and Control Systems

Hanum Shirotu Nida, Romie Oktavianus Bura, and Abdurahman

Abstract Information systems are needed to provide data and information anytime and anywhere to support the process of policy and decision making. In practice, Indonesia–Malaysia security border areas in West Borneo, each Indonesia military unit runs its own when monitoring the border without any interoperability or information sharing. Interoperability is needed to facilitate intermilitary unit of the information obtained by making it more efficient and effective for monitoring the Indonesia–Malaysia border in West Borneo, Indonesia. Also, intermilitary unit can be integrated with each other while facing the threats that might happen. The solution is to design a multi-network transmission system using socket programming as a communication media. Socket enables communication between one host and another using the same protocol. The protocol used in this application is TCP/IP because it has the advantage of guaranteeing data received its destination and no data duplication. The socket application model is based on a client–server which consists of one server and several clients. This is one of the advantages of a socket where the socket is able to handle multiple clients at the same time (multiple clients).

Keywords Socket programming · Data transmission · Network · Command and control · Interoperability

H. S. Nida (✉)
Indonesia Defense University, Bogor, Indonesia
e-mail: hanumshirotunida@gmail.com

R. O. Bura · Abdurahman
Bandung Institute of Technology, Bandung, Indonesia
e-mail: romiebura@idu.ac.id

Abdurahman
e-mail: abdurahmanfaqod@gmail.com

E. Joelianto et al. (eds.), *Cyber Physical, Computer and Automation System*, Advances in Intelligent Systems and Computing 1291,
https://doi.org/10.1007/978-981-33-4062-6_6

1 Introduction

The national defense information system has a role to support the implementation of national defense by providing fast, accurate data and information. The national defense information system must be able to provide data and information whenever and wherever to support the process of policy and decision making. The Indonesian National Armed Forces (TNI) information system is tasked with regulating the pattern of communication among the military unit (Army, Navy and Air Force) in order to organize a modern state defense. The integration of the military information system is called interoperability. Currently, the national defense information system is still not maximized because each unit has a different channel, so a standardization process is needed to integrate the information system. This is a challenge in developing a modern national defense because each military units, especially in West Borneo, Indonesia, uses different technology devices. In addition, the technology cannot be changed because they have their respective protection in accordance with the factory from which the technology was made, which is mostly made in foreign countries. In practice, securing the Indonesia–Malaysia border region in West Borneo, each unit runs on its own when conducting border monitoring without interoperability or information sharing. Information sharing is needed to facilitate inter-unit of the information obtained by making it more effective and efficient in monitoring the Indonesia–Malaysia border. Also inter-unit can be integrated with each other facing threats that might happen anytime and anywhere.

The interoperability development strategies that will be designed in this study are multi-network socket network transmission, based on client–server which has three clients, namely Regional Military Command (*Kodam*) XII Tanjung Pura, Main Base of Indonesia Navy (*Lantamal*) XII Pontianak, Command and Control Center (*Puskodal*) Airport-Based Supadio Kubu Raya and Mission Server. The headquarters of the Indonesian National Army (*Mabes TNI*) are shown in Fig. 1. Each client can communicate with each other through the server. The client acts as a transmitter and

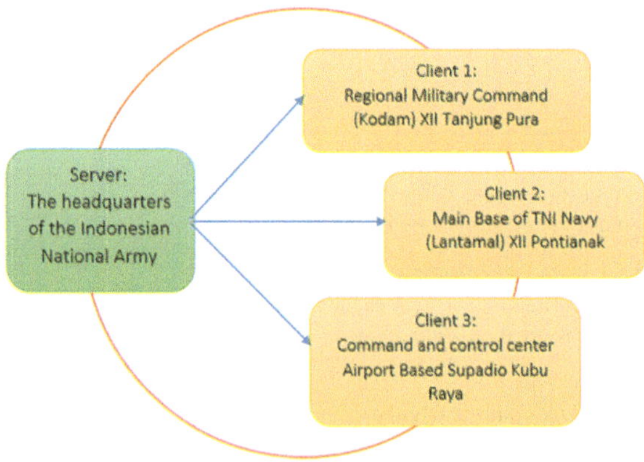

Fig. 1 Interoperability development strategy

receiver that have the authority to send and receive information sent by other clients through the server.

The implementation of interoperability in defense communication systems is in accordance with the availability of infrastructure and the urgency level of the interoperability network itself. The simplest solution is to standardize the existing communication system tools in each dimension.

2 Literature Review

2.1 TCP/IP

TCP/IP is one of the communication protocols that is used to solve a problem because protocol is designed in accordance with existing problems. The essential protocols in TCP/IP are Internet Protocol (IP), Transmission Control Protocol (TCP) and User Datagram Protocol (UDP). This research uses TCP protocol which provides reliable data transmission, full duplex operation (two-way data transfer), connection protocol, high application reliability and can be used to transfer large data [1]. In addition, the TCP protocol is designed to protect and recover lost data, data duplication and minimize other errors that may occur when sending data.

2.2 Socket Programming

Socket is a system that can allow an application to transfer and receive data, also read and write data into data storage. Socket makes an application that can be connected to the network and can be communicated with the other applications which are connected to the same network, so that information from one host can be read by applications on other hosts using the sample protocol and vice versa [2]. Examples of socket implementation are in client–server applications, cloud computing [3], web sockets [4], etc. In the C# programming language that uses the .NET framework, there are two specific classes for TCP socket called TcpClient and TcpListeners. Before starting a TCP connection, the host that acts as the client transmits a request message to the server. Technically, TcpListeners listens for TCP connection requests from client and creates a new socket in the TcpClient form to handle each incoming connection which transmitted to the server.

2.3 Client Server

The client–server model allows each computer to become a client or a server. Client is a program or application that allows users to access or view a program. Each client can be communicating and connecting with other clients through the application of the server. The concept is the client–server where the client is the system who started the transmission, while the server is a system that responds the requests from client passively [5]. The client–server program allows users to run client software to make queries. After client is connected, client transmits queries to the server. Server accepts client request and analyzes the query which is sent by the client and immediately transfers the answers to the client.

2.4 Interoperability

Interoperability is a concept about integrating an individual system into a system that is integrated with each other. The systems consist of the lowest level to the highest level in accordance with the components to be integrated. Interoperability includes standardization, cooperative integration and synergy between systems [6]. Interoperability determines connected components, including how each unit can exchange information and determine what decisions should be taken by the user. Data and information are sent in the form of text, images, videos, sounds, warning messages, sensor data, locations, etc.

3 Research Methodology

The research method used in this study is using waterfall models. Waterfall model is one of the sequential design processes that is often applied in the process of developing software where in development process flows like a waterfall. In the waterfall model, all requirements must be clear before moving to the next design phase so there is no overlapping in each phase [7]. Each phase is scheduled according to a predetermined time, and the testing process is to accomplish when the code is established completely. Figure 2 shows the flowchart of the waterfall model research design.

3.1 System Analysis

Technically, the data is converted into cipher text and then transmitted through the applications layer, TCP sockets, TCP ports, TCP/UDP (in this study using the

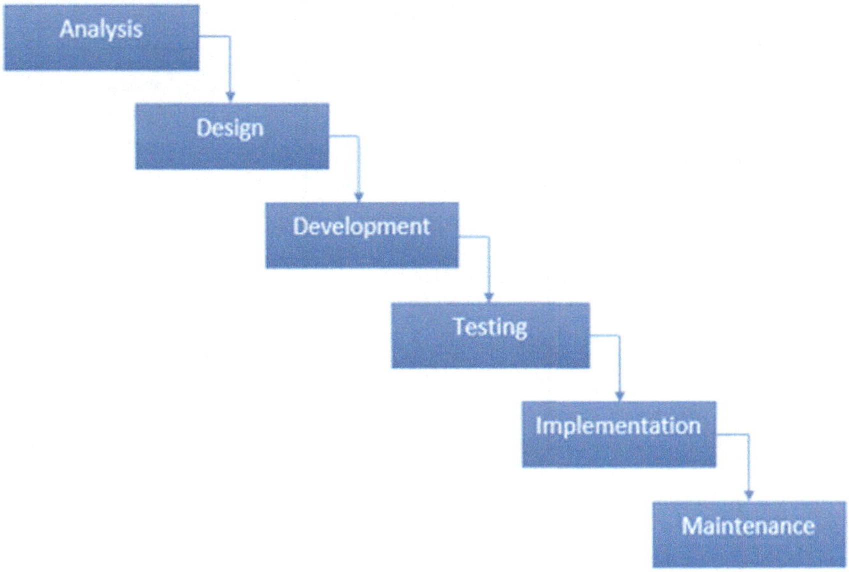

Fig. 2 Waterfall model life cycle

protocol TCP), IP, channel (such as Ethernet), routers, channel, IP, TCP, TCP ports, TCP socket, application layer and then ciphertext are decrypted so that they are the initial plain text that is shown in Fig. 3.

This study uses TCP because it has several advantages over UDP, and TCP guarantees data sent to destination addresses safely because TCP is designed to detecting and recovering lost data in the middle of the transmission. TCP overcomes data duplication and other problems that may occur when data is sent across a network [8]. The method used by TCP to guarantee reliable delivery is to use the concept of confirmation and retransmission. The sender of the packet keeps track of every transmitted packet and waits for approval of successful delivery from recipient before transmitting the next package. The sender makes an internal timer where the packet will be automatically resent if at the specified time defined by the sender has not received confirmation [9]. TCP uses the serial number on each packet as a confirmation message to ensure the packet has been sent to the recipient, so there is no duplicated data when delay occurs during transmitting data. The connection is done by completing a handshake message on the TCP application between the two computers that are communicating [10].

The advantage of socket programming is that there is no absolute requirement for ports to be used in transmitting data. There are many port options that can be used and not interfered with other running process in a system. The port is a virtual destination that allows nodes to carry out multiple network communication simultaneously. Conceptually, the port is assumed to be a door through which information enters and exits a node. For example in the Linux system, there are 65,535 ports that can be

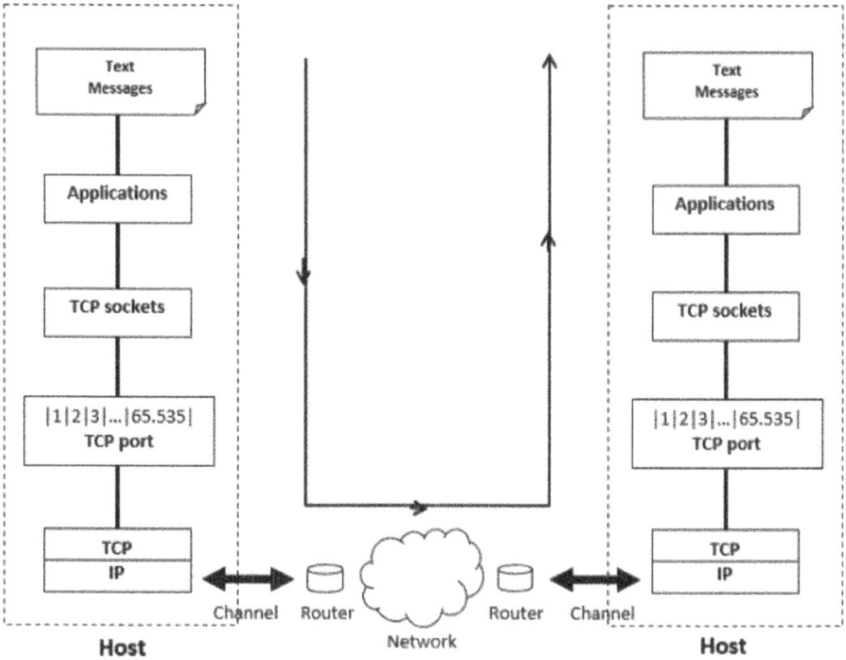

Fig. 3 System flowchart

available, but there are several ports that have been used and cannot be changed such as port 21 for FTP, port 23 for telnet, port 80 for HTTP, etc. The format of writing address is to use an IP address and the port number, for example 192.168.1.2:43, which means port 443 at address 192.168.1.2.

3.2 System Design

This research is using client/server design which consists of one server and three clients that are shown in Fig. 4.

Server flowchart shown in Fig. 5 (left), illustrating how the system works, after the server activated the user enters the user name and password as authentication. If the values of username and password entered are true, the system will enter the main page and automatically listen to the client. When the client tries to connect the server, the server will have choice to accept client or not. If server selects yes, then the client will be linked to the server, if not then the client cannot connect to the server. The server can receive messages from clients and can send messages to selected clients. Client flowchart is shown in Fig. 5 (right) which illustrates the client system works where the client must log into the system first as authentication. After that the client can make a request to connect to the server, if the server receives a request from

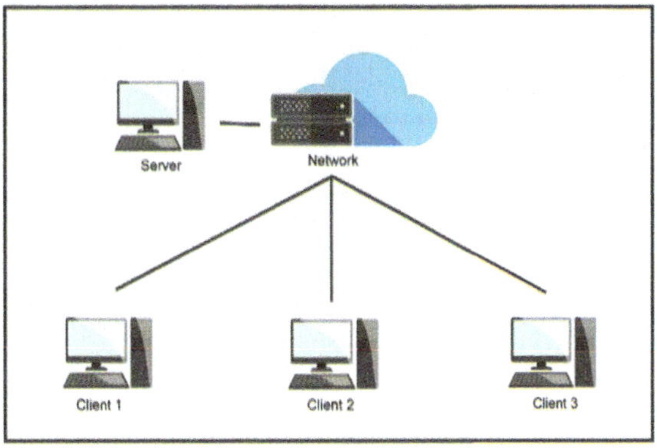

Fig. 4 Client/server design

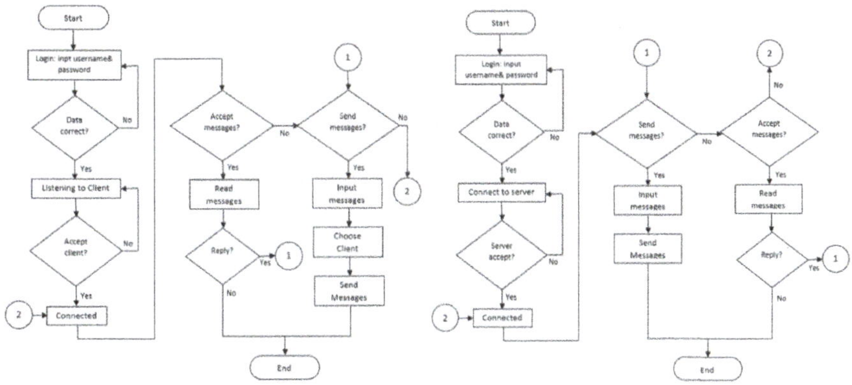

Fig. 5 Server and client flowchart

client then client can be connected to the server. After the client is associated with the server, the client can transmit and receive messages from the server.

3.3 System Development

This program was developed using C# programming language and socket system. Multi-socket network transmission system is designed to have several functions such as login, connect to server, send messages and log out. Figure 6 shows the server and client design, both of the program could send and receive messages.

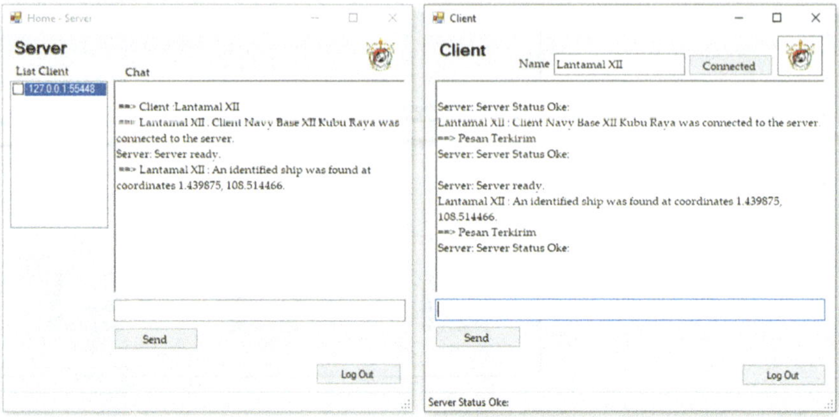

Fig. 6 Server and client design

3.4 *Testing*

Black box testing was conducted in this study to test the functionality of the system and ensure that the function can run according to the design of the system. Black box testing is done by carrying out procedures or testing steps to find out whether the test results are in line with expectations or not. Black box testing is described in Table 1.

3.5 *Implementation*

This system can be implemented using a virtual private network (VPN) to ensure the security of a network. To implement the system in one region of West Province, this system can use a dial-up or cable Internet connection from a commercial Internet service provider (ISP) thought it could be reached. Another option besides using an ISP is to use a broadband network such as Internet satellite. Each node must have a very small aperture terminal (VSAT) to transmit data to satellites. The use of this broadband network has many advantages, namely having a high data transfer rate and always active so that it can be used in an emergency such as when a disaster occurs where the Internet network is broken and cannot be used.

4 Conclusion and Future Work

This research is needed to actualize the interoperability communication system and answer the communication needs of the Indonesian National Army especially to

Table 1 Black box testing

Test ID	Test description	Steps	Expected Result	Result
A.01	Login function	Input the correct username and password	The message box "Success login" appears and windows move to the home	Valid
A.02	Login function	Input incorrect username and password	The message box "Wrong username and password" appears	Valid
A.03	Exit button	Click the exit button	Exit the windows application	Valid
A.04	Client start connection to the server	Enter the client's name and click the "Connect to Server" button	The client gets a "Connected" message and the client can send a message to the server	Valid
A.05	Client sending messages to the server	Type a message and click the send button	The server received messages from the client and the message appears in the message fields	Valid
A.06	Server sending messages to the client	Type the message and choose the destination client	The client receives the message and the message appears in messages fields	Valid
A.07	Log out function	Click the log out button	The message box "Do you want to log out?" appears, if yes then the windows application will be closed.	Valid

integrate units in West Borneo Province so that the command and control system of the Indonesian National Army which is a basic need can work properly. This research presents the implementation of socket programming in multi-networks transmission using .NET framework and waterfall method. With this research, new discoveries related to transmit data using socket programming can be deeper discussed as well as the advantages and disadvantages of the system so that it can be a reference in the future. The researcher recommends further research on socket programming security such as encryption for the security of data transmitted over a network.

Acknowledgements This work is supported by the Capacity Building Program of the Faculty of Defense Technology, Indonesia Defense University.

References

1. Zhang, B., Xia, Q., Han F.; Multi-point-to-point image transmission system based on TCP/IP protocol. In: 2015 fifth international conference on instrumentation and measurement, computer, communication and control (IMCCC), Qinhuangdao, pp. 1621–1624 (2015)
2. Maata, R.L.R., Cordova, R., Sudramurthy, B., Halibas, A.: Design and Implementation of client-server based application using socket programming in a distributed computing environment. In: 2017 IEEE international conference on computational intelligence and computing research (ICCIC), Coimbatore, pp. 1–4 (2017)
3. Prajapati Ashishkumar, B., Barkha, P.: Implementation of DNA cryptography in cloud computing and using socket programming. In: 2016 international conference on computer communication and informatics (ICCCI), Coimbatore, pp. 1–6 (2016)
4. Liu, Q., Sun, X.: Research of web real-time communication based on web socket. Int. J. Commun. Netw. Syst. Sci. **05**, 797–801 (2012)
5. Xue, M., Zhu, C.: The socket programming and software design for communication based on client/server. In: 2009 Pasific-Asia conference on circuits, communications and system, pp. 775–777 (2009)
6. Sumerta, Gede, I., et al.: Information System of the Indonesian National Armed Forces within the Interoperability Data Link State Defense. Indonesia Defense University (2017)
7. Balaji, S., Sundararajan M.: Waterfall versus V-Model versus agile: a comparative study on SDLC. Int. J. Inf. Tech. Bus. Manag. **2**(1) 2012
8. Davis, K., Turner, J.W., Yocom, Nathan: The definitive guide to linux network programming, pp. 3–39. Springer, New York (2004)
9. Donahoo, Michael J., Calvert, Kenneth L.: TCP/IP SOCKET in C: Practical Guide for Programmers Second Edition, pp. 1–8. Elsevier, San Francisco (2009)
10. Makofske, D.B., Donahoo, M.J., Calvert, K.L.: TCP/IP Socket in C#: practical guide for programmers, 2nd edn, pp. 15–17. Elsevier, San Francisco (2001)

Control Prototype of Manipulator Robot for Skin Cancer Therapy

Dessy Novita, Andri Abdurrochman, and Asep Sholahuddin

Abstract The treatment (therapy) in clinical medicine needs highly precision positioning to localize the disease. In general, a clinician can locate and treat the disease manually. Some spots may be missed in practice. Therefore, we propose a manipulator robot for skin cancer therapy which can be equipped with a laser system to disintegrate unwanted tissues on the patient's skin. The main feature of this robot is the capability to scan the whole body's skin to localize the skin cancer by driving the manipulator in circular or elliptical skimming. Our prototype is a model of six joints manipulator robot and controlled by the PID control system for the servo dynamixel that optimized by extremum seeking control based on simulation. This combination results a faster response to the unity step input.

Keywords Prototype manipulator robot · Servo dynamixel · PID control · Extremum seeking control

1 Introduction

The development of robotics today is more advanced and widely used to assist human work. Research in robotics is either mobile robots, arm robots (manipulators) or combined mobile manipulators and humanoid robots designed like humans. In robotics, there are some application of mechanics, electricity, electronics, control, sensors, interfaces, actuators, software, artificial intelligence and others [1].

D. Novita (✉)
Department of Electrical Engineering, Universitas Padjadjaran, Bandung, Indonesia
e-mail: d.novita@unpad.ac.id

A. Abdurrochman
Department of Physics, Universitas Padjadjaran, Bandung, Indonesia

A. Sholahuddin
Department of Computer Science, Universitas Padjadjaran, Bandung, Indonesia

E. Joelianto et al. (eds.), *Cyber Physical, Computer and Automation System*, Advances in Intelligent Systems and Computing 1291,
https://doi.org/10.1007/978-981-33-4062-6_7

There are also some researches on robotics for applications in medical and health care. Some worth to mention are: The da Vinci robot which applied intuitive surgical for the minimally invasive surgery [2] and Robot-assisted pelvic lymphnode dissection (RAPLND) at Guy's Hospital for skin malignancy. The RAPLND was designed for carcinoma of Merkel cell, which was multidisciplinary supervised and also employing a four-arm da Vinci robot [3]. Robotic pelvic lymphadenectomy with convenient vision and minimum morbidity can be treating patients with metastatic melanoma including the pelvic lymph nodes as safely [4]. Some proposed manipulator robots as supporting tools for minimally invasive surgery with new procedure, kinematic structure, design and construction [5]. Including, continuum robots to improve the rigidity of links manipulator which will enhanced performanced robots for medical applications [6], or adding automatic gain controller by extremum seeking control [7–10].

As robots can be operated as desired to any location accurately, including for medical application, the robot manipulators may have a high precision positioning. For cancer therapy, it is must to have the high precision positioning directing the laser to the right location and destroys the cancer on the skin. This paper reports how we build the robot arm or manipulator for skin cancer therapy. It is capable of operating tools and identified the cancer location by moving the scanner around the patients then positioning the laser manipulation to the exact location for the therapy.

Skin cancer is classified in malignant melanoma and non-melanoma; that has two main types, basal and squamous cell carcinoma [11–13]. Malignant melanoma is a skin cancer that is hard to predict in advanced stage. Most skin cancers are non-melanoma. Basal cell carcinoma is a general skin cancer that is originated at the basal layer in the epidermis and commonly does not show any preliminary lesions. The higher the growth and the side effects of existing treatments, attract any alternative therapy research combining photodynamic therapy applying low-intensity laser. Low-intensity laser therapy does not cause cellular activity and is used as a major treatment by inducing chemical reactions into the tissues based on long periods of low-intensity light exposure [14]. This therapeutic technique can be applied to several molecular light absorption clinic fields and can activate the movement conditions. The optical radiation used in the treatments is a 633 nm HeNe laser. The effect depends on the amount of energy applied and the power density source preferably according to the desired effect. The density energy carries its own effect: anti-inflammatory from 1 to 6 J/cm^2; eutrophic from 3 to 6 J/cm^2; circulatory from 1 to 3 J/cm^2 and antialgic from 2 to 4 J/cm^2 laser [15].

The main purpose of this paper is reporting our manipulator robot for skin cancer therapy with applied design parameters is based on reference sources of the P560 dynamic model of the P560 dynamic model which is taken from the reference model of Peter I. Corke [1, 16, 17]. The model and control manipulator is analyzed by dynamic and robust control for flexible link manipulator that is described in [18] and kinematic control of redundant with multiple constraints in [19].

2 Methodology

2.1 Modeling of Manipulator Robot

To create a prototype of a robot manipulator, it is necessary to design the robot model and to analyze the robot's parameters. Denavit–Hartenberg (D-H) parameters are used for describing link relations which are assumed rigid body. Denavit–Hartenberg parameters consist of link length (a_i), link twist (α_i), link offset (d_i) and joint angle (θ_i) [1]. The direct kinematics model is built using the D-H transformation matrix representation, and they are the coordinate connection of the robot manipulator with six joints. Transformation matrix arm with six joints is described from forward kinematic on Eq. (1) [1].

$$T_6 = \begin{bmatrix} n_x & s_x & a_x & p_x \\ n_y & s_y & a_y & p_y \\ n_z & s_z & a_z & p_z \\ 0 & 0 & 0 & 1 \end{bmatrix} = {}^0A_1 * {}^1A_2 * {}^2A_3 * {}^3A_4 * {}^4A_5 * {}^5A_6 \quad (1)$$

Dynamic analysis described the dynamic behavior of the manipulator robot by attention to the forces that caused the movement. The model of dynamic robot manipulator is shown on Eq. (2) [1].

$$\tau(t) = H(q(t))\ddot{q}(t) + h(q(t)\dot{q}(t)) + G(q(t)) + \tau_Q \quad (2)$$

where $\tau(t)$ is vector of torque on actuator each joint i,

$$\tau(t) = (\tau_1(t), \tau_2(t), ..., \tau_n(t))^T, \quad i = 1, 2, 3, \ldots, n \quad (3)$$

$H(q)$ is vector of transformation dynamic matrix ($n \times n$), n is n-DOF of manipulator robot, $h(q)$ vector of torque matrix ($n \times 1$) Coriolis effect and centrifugal motion on joint i,

$$h(q, \dot{q}) = (h_1, h_2, ..., h_n)^T, \quad i = 1, 2, 3, \ldots, n \quad (4)$$

$$h_i = \sum_{k=1}^{i} \sum_{m=1}^{i} h_{ikm} \dot{q}_k \dot{q}_m, \quad i, k, m = 1, 2, 3, \ldots, n \quad (5)$$

$G(q)$ is vector of torque matrix ($n \times 1$) gravitation effect,

$$G(q) = (g_1, g_2, \ldots, g_n)^T, \quad i = 1, 2, 3, \ldots, n \quad (6)$$

τ_Q is disturbance or load vectors [1].

2.2 PID Control Optimized by Extremum Seeking Control

The robot manipulator is controlled in each joint by the servomotor. The control method is proportional, integral and derivative (PID) control. PID control can be operated in every joint and optimized by extremum seeking control (ESC) that is proposed. The block diagram of PID control on each joint of robot and optimized by ESC is shown in Fig. 1. The block diagram of ESC algorithm is depicted on Fig. 2. Before updating PID settings, PID control requires initial parameters that should be a stable system.

Parameters of PID controller *C(s)* are proportional gain (K_p), integral constant (K_i) and derivative constant (K_d). Parameters of motor DC for simulation are applied which consist of electromotive force constant (*Ka*: Nm/Amp ($K = K_e = K_t$)), electric resistance (*Rs:* ohm), electric inductance (*Ls*: Henry), moment of inertia (J_{eff} kg m^2/s^2), damping ratio of the mechanical system (f_{eff}: Nms), ratio of gear (n), motor DC constant (K_b), desired angle (θ_d), actual angle (θ_a) and cost function ($J(\theta)$) shown in Fig. 1 and Fig. 2. The control variable is the angle of joint that gets error from discrepancy between desired angle (θ_d) and actual angle (θ_a) while the cost function calculated integral of square error. The extremum seeking algorithm has the first step gradient estimation that utilizes high-pass filter, low-pass filter and perturbation signal ($\alpha_j \cos(\omega_j k)$), and then the second step is updating variables for the minimum cost function by:

$$\theta(k+1) = \theta(k) - \gamma \nabla J(\theta(k)) \tag{7}$$

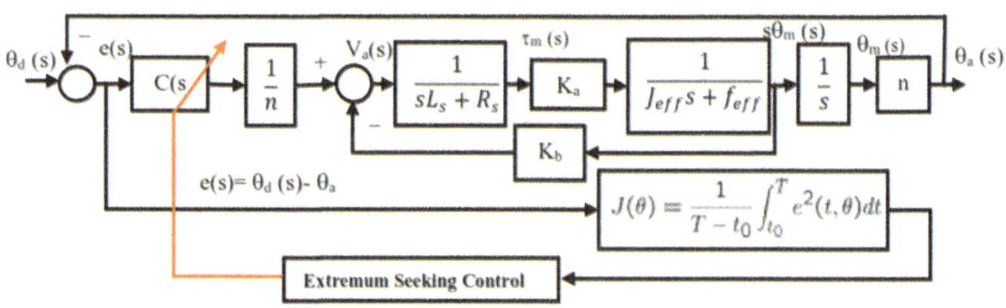

Fig. 1 Diagram block of PID control each joint and optimized by extremum seeking control

Fig. 2 Block diagram of algorithm extremum seeking control [10]

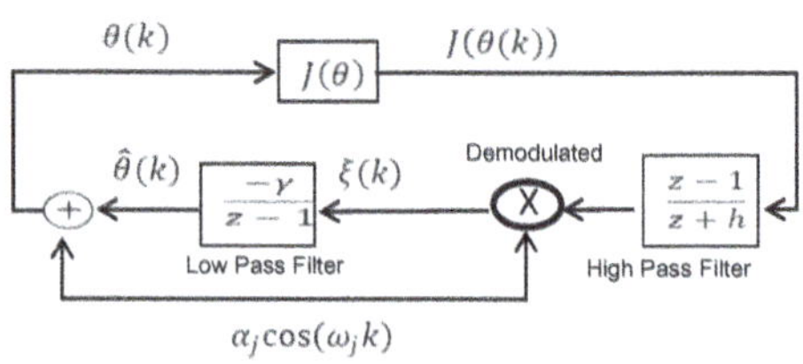

where θ is the angle of joint, γ is updating step, and $\nabla J(\theta)$ is gradient cost function [7–10].

The procedure of this research is described as follows.

1. Design robot arm or manipulator.
2. Look for Denavit–Hartenberg parameters and simulate for kinematic and dynamic analysis.
3. Selection of motor according to torque from result of dynamic and kinematic analysis.
4. Servomotors in each joint are simulation and testing. The Dynamixel Wizard is used for control of moving servomotors each joint.
5. Design of control methods, programs and simulations. Joints are applied by PID control and optimized by extremum seeking control.
6. Mechanical hardware robot arm, actuator and controller with laboratory work.

3 Results and Analysis

Design of therapy robot on this research is shown in Fig. 3. It has six joints robot manipulator. It can be seen in Fig. 4 the prototype of manipulator robot. The robot has degree of freedom (DOF) on Table 1 and also the direction motion on each joint. Forward kinematic model of manipulator robot is created using Denavit–Hartenberg representation of transformation matrix which described six joints on Table 2 for relationships between adjacent links as the translational and rotational motions.

The parameters of manipulator robot for medical instrument consist of:

- The mass for each of the nth arm can be seen from Table 2.
- The position of the arm center i from the center of the baseline frame (r_i) is shown in Table 3.

Moreover, to analyze the design of this manipulator robot, the selection of the link material and servomotors is rotational motion test is conducted for each servo

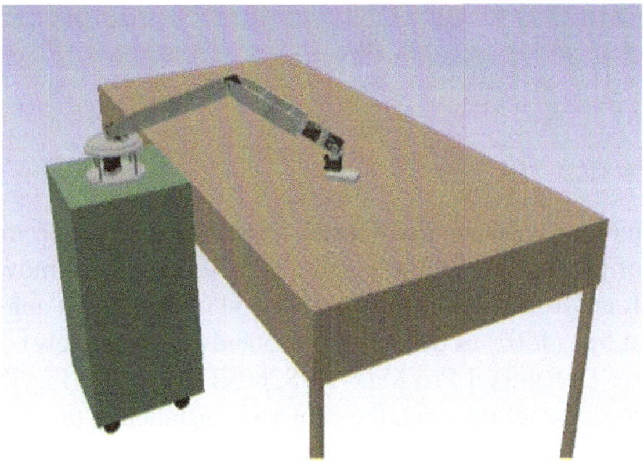

Fig. 3 Design of therapy robot

Fig. 4 Prototype of manipulator robot

Table 1 Degree of freedom (DOF) and servomotor each joint

Joint	DOF	The direction of motion	Dynamixel servo motor
1	1	Left–right	RX-64
2	1	Above-below	Dynamixel Pro M42-10-S260-R
3	1	Above-below	EX-106
4	1	Left–right	RX-28
5	1	Above-below	RX-28
6	1	Wrist or rotating	RX-28

Table 2 Parameters of D-H robot manipulator six joints, mass for nth arm

Joint	Joint angle $\Theta°$	Link twist $\alpha°$	Link length a (mm)	Link offset d (mm)	Joint range	Mass (kg)
1	0	0	0	152	0°–300°	0.648
2	90	90	0	70	150°–300°	0.839
3	0	0	500	0	0°–300°	1.100
4	90	90	300	0	50°–200°	0.746
5	90	90	72.6	0	50°–200°	0.744
6	90	90	72.6	0	0°–300°	2.012

motor by dynamixel wizard, hence the joint angle may set as desired. The setting of joints angle is 150°, and the servomotors are moved according to the set angle. Joints 1, 3, 4 and 5 are rotated at 511 (the current angle position) with goal position at 512 (150°) as desired (Figs. 5 and 7, respectively). The test of Dynamixel Wizard for Dynamixel Pro M42-10-S260-R Joint 2 is (Fig. 6) at 0° as goal position. Joint 6 can be tested when the camera is installed at the end effector.

Table 3 Position of the arm center (i) from the center of the baseline frame (r_i)

Parameters of position	Value (mm)
S_{x1}	0
S_{y1}	0
S_{z1}	75
S_{x2}	0
S_{y2}	0
S_{z2}	145
S_{x3}	500
S_{y3}	0
S_{z3}	0
S_{x4}	300
S_{y4}	0
S_{z4}	0
S_{x5}	72.6
S_{y5}	0
S_{z5}	0
S_{x6}	0
S_{y6}	0
S_{z6}	72.6

The simulation PID control by optimized ESC which has the input as unity step function. The initial PID parameters according to Routh–Hurwitzh is $K_p < 0.32$, and so we set $K_p = 0.1$, $K_i = 0$, $K_d = 0$. In simulation, the motor DC parameters are $K_a = 7.67\text{E-3}$ (Nm/Amp: electromotive force constant ($K = K_e = K_t$)), $Rs = 2.6$ (ohm: electric resistance), $Ls = 1.8\text{E-6}$ (Henry: electric inductance), $J_{eff} = 5.3\text{E-7}$ (kg m^2/s^2: moment of inertia of the rotor), $f_{eff} = 7.7\text{E-6}$ (Nms: damping ratio of the mechanical system), $n = 10$ (ratio of gear), $K_b = 7.67\text{E-3}$ (0.0274;) and $T_d = 0$ (load disturbance) [1]. The simulations result in Fig. 8 shows the system response to the step function with (a) PID control and (b) PID optimized by ESC. The optimization results are updated for each ten iterations with the updating step $\gamma = 0.001$. The PID parameters are also updated to $K_p = 0.1$, $K_i = -6.28788566510961\text{e-}07$ and $K_d = -2.42120924944676\text{e-}07$.

4 Conclusions

The simulation of angle joint applied PID control and ESC optimizing system demonstrates faster response to the unity step input. The designed prototype robot manipulator is worked out. The construction is regarding to the robot framework, mass, link

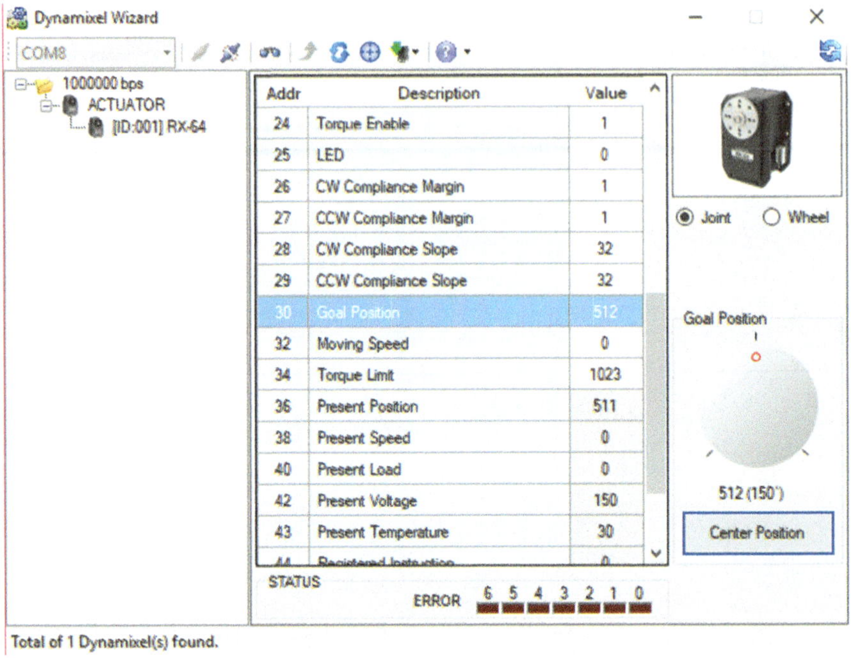

Fig. 5 Test of dynamixel wizard of RX 64 Joint 1

length, motor torque as desired. Nevertheless, this prototype needs further development to be a fully operating skin cancer therapy apparatus that equipped with artificial intelligent, image processing of skin cancer, optimized by extremum seeking control and LASER for skin cancer treatment.

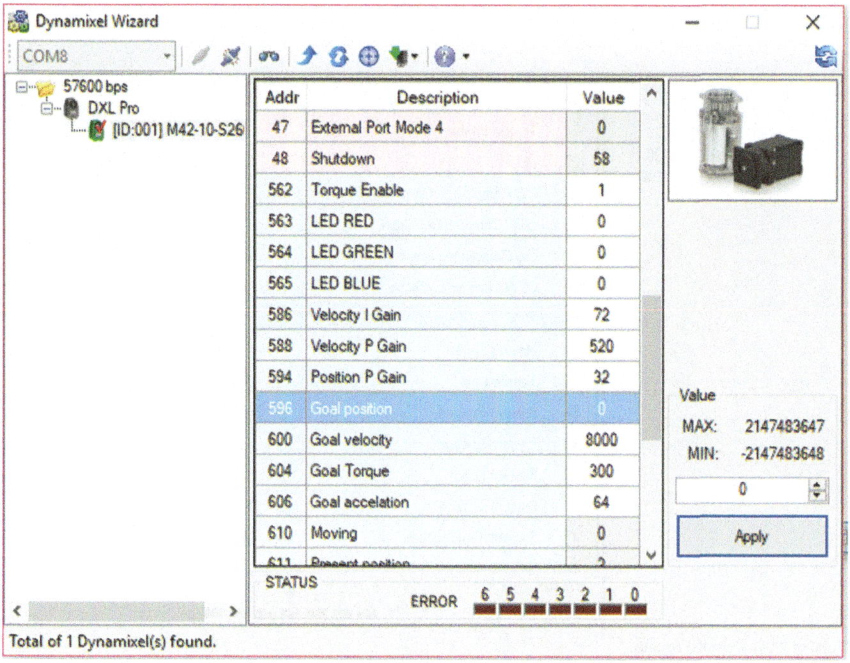

Fig. 6 Test of dynamixel wizard of dynamixel Pro M42-10-S260-R Joint 2

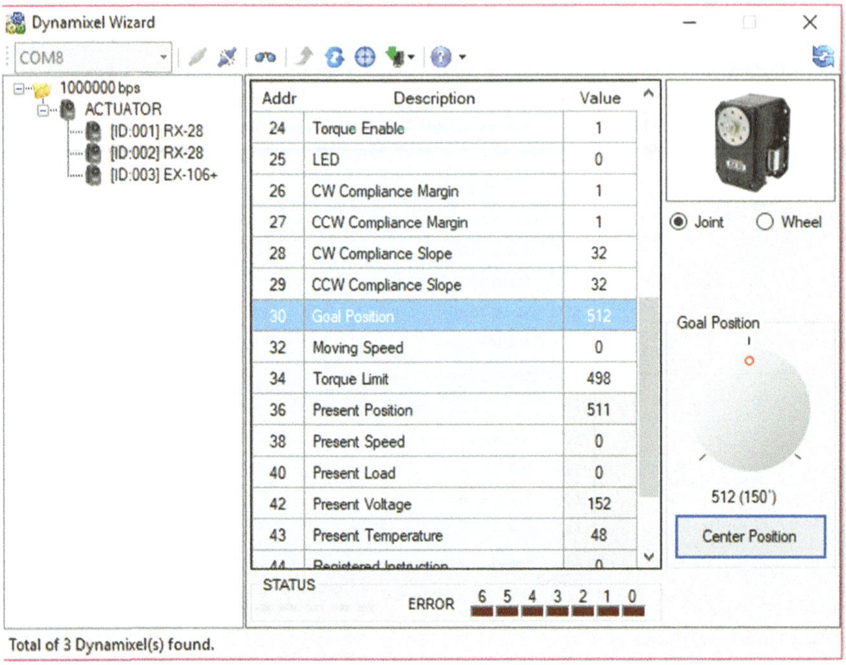

Fig. 7 Test of dynamixel wizard of RX 28 for Joint 3, 4 and 5

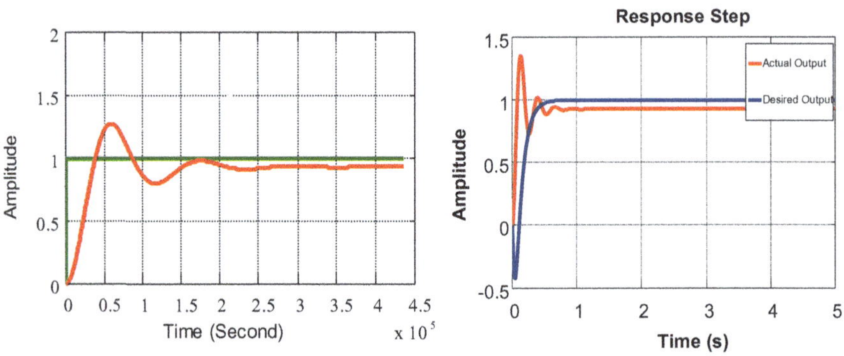

(a) Green line : desired output,
Red line : actual output

(b)

Fig. 8 Step response of the robot joint system **a** PID control and **b** PID optimized by ESC

Acknowledgements The authors would like to express gratitude for Universitas Padjadjaran, Rector of UNPAD, Directorate of Research, Community Service and Innovation of Universitas Padjadjaran, Ministry of Research, Technology and Higher Education of the Republic of Indonesia (RISTEKDIKTI), for research funding of internal UNPAD, Department of Electrical Engineering,

Universitas Padjadjaran, fellow research teams and students who participated in this research at Electrotechnic, Electric Power and Communication Technology Laboratory.

References

1. Fu, K.S., Gonzalez, R.C., Lee, C.S.G.: Robotics Control, Sensing, Vision, and Intelligence. McGraw-Hill, New York (1987)
2. Harvey, C., Dragone, R.: Surgical Robotics: The Evolution of a Medical Technology. In: Medicaldesignbriefs.com. (2016). (Online). https://www.medicaldesignbriefs.com/component/content/article/mdb/features/articles/25006?start=1. Accessed: Nov 13 2019
3. MacKenzie Ross, A.D., Kumar, P., Challacombe, B.J., Dasgupta, P., Geh, J.L.C.: The addition of the surgical robot to skin cancer management. Annal. Roy. Coll. Surg. Engl. **95**, 70–72 (2013)
4. William, S., David, S.F., James, J., David, K.O.: Robot-assisted laparoscopic transperitoneal pelvic lymphadenectomy and metastasectomy for melanoma: initial report of two cases. J. Robot. Surg. **4**, 129–132 (2010)
5. Mianowski, K.: Design study of mechanical robot-manipulators for medical applications. MeSRob (2016)
6. Burgner-Kahrs, J., Rucker, D.C., Choset, H.: Continuum robots for medical applications: a survey. IEEE Trans. Robot. **31**(6), 1261–1280 (2015)
7. Dessy, N., Shigeru, Y.: Extremum seeking for dead-zone compensation and its application to a two-wheeled robot. J. Auto. Control Eng. **3**(4), 265–269 (2015). https://doi.org/10.12720/joace.3.4.265-269https://doi.org/10.12720/joace.3.4.265-269
8. Chan, M., Kong, K., Tomizuka, M.: Automatic controller gain tuning of a multiple joint robot based on modified extremum seeking control. In: The 18th IFAC World Congress, Milano (Italy), pp. 4131–4136, 28 Aug–2 Sep (2011).
9. Killingsworth, N.J., Krstic, M.: PID tuning using extremum seeking. In: IEEE Control System Magazine, pp. 70–79 (2006)
10. Kong, K., Inaba, K., Tomizuka, M.: Real-time nonlinear programming by amplitude modulation. In: Proceedings of DSCC2008 ASME Dynamic Systems and Control Conference, Ann Arbor, Michigan, USA, 1–8, 20–22 Oct (2008)
11. Gordon, R.: Skin cancer: an overview of epidemiology and risk factors. Semin. Oncol. Nurs. **9**, 160–169 (2013)
12. Lazareth, V.: Management of non-melanoma skin cancer. Semi. Oncol. Nurs. **29**, 182–194 (2013)
13. Almansour, E., Jaffar, M.A.: Classification of dermoscopic skin cancer images using color and hybrid texture features. IJCSNS Int. J. Comput. Sci. Netw. Secur. **16**(4) (2016)
14. Fanjul-Velez, F., Salas-Garcia, I., Torre-Celeizabal, C., Zverev, M., Aree-Diego, J.L.: Analysis of low intensity laser therapy as adjuvant to photodynamic therapy in nonmelanoma skin cancer. IEEE Explore 978–1–4244–9270–1/15 (2015)
15. Karu, T.I.: Ten Lectures on Basic Science of Laser Phototherapy. Prima Books AB (2007)
16. Corke, I.P., Armstrong, B.: Journal of A Search for Consensus Among Model Parameters Reported for the PUMA 560. Australia.
17. Otter, M.: Journal of Modelica A Language for Physical Modeling, Visualization and Interaction. Germany, August, (1999).
18. Mohammad, K., Zaharuddin, M., Abdul, R.H.: Dynamic model and robust control of flexible link robot manipulator. TELKOMNIKA **9**(2), 279–286 (2011)
19. Shuihua, H., Ji, X., Wei, W., Michael, Z.Q.C.: On the virtual joints for kinematic control of redundant manipulators with multiple constraints. IEEE Trans. Control Syst. Technol. **26**(1), 65–76 (2018)

Video-Based Container Tracking System Using Deep Learning

Basuki Rahmat, Yisti Vita Via, Abdullah Wasian, Intan Yuniar Purbasari, Ni Ketut Sari, Widi Wurjani, Steven Bandong, Endra Joelianto, and Persaulian Siregar

Abstract The process of loading and unloading containers at the port is very important to be automated in order to increase productivity, revenues, efficiency and safety in the logistics transportation process especially in a maritime country like Indonesia. To achieve this, identification and tracking of container positions need to be done accurately so that container transfers can occur precisely and smoothly. In this study, YOLO deep learning is used to detect moving containers. The model is trained using container images, and then the validation and testing process are carried out on the model using other container images. The results of training are stored in several checkpoints which will be compared to get the most accurate model. Obtained YOLO gave good results for tracking containers with MAP value of 68.63% and LAMR of 0.31.

Keywords Port automation · Container tracking · Deep learning

1 Introduction

A container is a small warehouse that runs with the help of vehicles from one place to another, or from one country to another. Some of the benefits of containers include the following: benefits for shippers: reducing transportation costs, save port costs,

B. Rahmat · Y. V. Via · A. Wasian · I. Y. Purbasari · N. K. Sari · W. Wurjani
Faculty of Computer Science, Universitas Pembangunan Nasional "Veteran" Jawa Timur, Surabaya, Indonesia
e-mail: basukirahmat.if@upnjatim.ac.id

S. Bandong
Engineering Physics Doctoral Program, Faculty of Industrial Technology, Institut Teknologi Bandung, Bandung, Indonesia

E. Joelianto (✉) · P. Siregar
Instrumentation and Control Research Group, Faculty of Industrial Technology, Institut Teknologi Bandung, Bandung, Indonesia
e-mail: ejoel@tf.itb.ac.id

E. Joelianto et al. (eds.), *Cyber Physical, Computer and Automation System*, Advances in Intelligent Systems and Computing 1291,
https://doi.org/10.1007/978-981-33-4062-6_8

reducing warehousing and inventory costs, reducing packing costs, reducing insurance premiums, more comfortable, easier and better reception (port), the emergence of new markets, benefits for boat owners: speed up completion time, more cargo-carrying capacity, high return on investment, global contract, higher profitability, and inland operations, and benefits for port authorities: reducing port congestion, saving time, load and unload quickly and comfortably, fewer marketing efforts, and rationalization of cargo handling costs. So, containers are needed for export and import activities to save costs, time, easier processes and so on.

Realizing the importance of containers for export and import activities, many attempts have been made to automatically unload and load containers. Some research on the container system automatically can be mentioned, among others [1–4]. In the process of loading and unloading the container system automatically, it can only be done if the system can identify and recognize the position of the container. Colors and types of container are general standards for marking containers to remain in position so they can easily record data on the container. Some research using R-CNN [5], Fast-RCNN [6], SSD [7] and YOLO [8] had been implemented in object detection area. The YOLO network model has good performance in real-time object detection [9]. Therefore, this study proposes video-based container tracking using YOLO.

2 Deep Tracking

The application of deep neural network to tracking object is referred as deep tracking. One method which implements deep learning for tracking is the convolutional neural network (CNN) method. It is a special type of neural network for processing data that has a grid topology like image. Convolution operations are carried out at least in one layer in CNN. This method is known as appropriate model for solving object recognition and detection problems. Technically, CNN consists of several stages, namely the convolution layer, the pooling layer and the fully connected layer [10].

2.1 Convolutional Layer

Convolutional layer is the first layer that accepts input images directly to the architecture and does convolution operation. Feature extraction from input images is the main aim of convolution process. Equation below shows the formula to do convolution [11].

$$s(t) = (x * w)(t) \tag{1}$$

Equation (1) creates a single output called feature map, the x is the input, and w is known as kernel or filter. If the input is two dimension like image, then (t) can be thought as a pixel and replaced by i and j. Therefore, for convolution operations with more than one dimension, the following equation can be used [11]:

$$S_{(i,j)} = (K * I)_{(i,j)} = \sum \sum I_{(i-m,j-n)} K_{(m,n)} \tag{2}$$

Equation (2) is a convolutional computation with pixels of an image represented by i and j. K is the kernel, and I is the input.

2.2 *Pooling Layer*

Pooling is a matrix size reduction using the pooling operation. Pooling layer takes place after convolutional layer. There are two kinds of pooling that are often used, namely average pooling and max pooling. Average pooling takes the average value, but at max pooling, the maximum value. The output of the pooling process is a matrix with smaller dimensions compared to the initial matrix. The convolution and pooling process is carried out several times so that a feature map of the desired size is obtained. The feature map will be an input for fully connected neural networks [12].

2.3 *Fully Connected Layer*

The results of the convolutional layer and pooling layer are still in the form of multidimensional arrays. It needs to be reshaped into a one-dimensional vector so that it can be used as input for the fully connected layer. This layer has a hidden layer, activation function, output layer and loss function. The result of this layer is the decision of what class the image belongs to [13].

One method of object detection that applies CNN is the You Only Look Once (YOLO) method. YOLO applies a single neural network to the whole picture. This network will divide the image into regions and then predict the boundary box and its probabilities, for each boundary area box the probability is weighted to classify as objects or not. In this study, we used Tiny YOLO which consists of seven convolutional layers, six pooling layers and two fully connected layers (Fig. 1). Filters start from 16, 32, 64 and so on (multiplied by 2). Filters size for convolutional layers is 3 × 3, while for pooling layers is 2 × 2 with a stride of 1. Each output from pooling will be continued as input for the next convolution.

This method breaks the input image into S × S grid cells, predicts the bounding box B for every cell and calculates scores for each class C. There are five predictions of the bounding box which are center x, center y, width, height and confidence level of the bounding box. All bounding boxes generated in each grid cell will form a class score set C. Therefore, for each image processed by YOLO will give output as vector S × S × (5B + C). The extracted features from the convolutional layer then

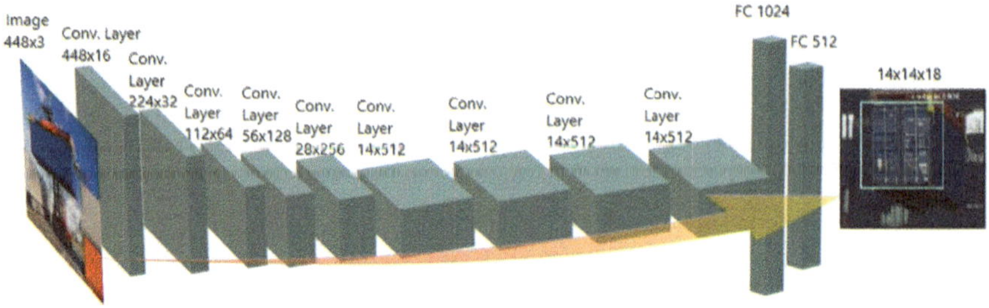

Fig. 1 Tiny YOLO architecture

transferred to fully connected layer that predicts object probability and constructs the bounding box. The result of the fully connected layer will be regressed by the final detection layer that maps it to the corresponding bounding box and class [14]. Figure 1 shows the Tiny YOLO architecture.

The accuracy of the object detection results is calculated using mean average precision (MAP) and the log average miss rate (LAMR). The MAP gives the form of the precision/recall curve. MAP is the average precision of eleven uniformly spaced recall levels (0, 0.1, ..., 1) [15]:

$$\mathrm{MAP} = \frac{1}{11} \sum_{r \in \{0, 0.1, \ldots, 1\}} p_{\mathrm{interp}(r)} \tag{3}$$

where

$$p_{\mathrm{interp}(r)} = \max_{\tilde{r}:\tilde{r} \geq r} pr \tag{4}$$

$$pr = \frac{tp}{tp + fp} \tag{5}$$

$$re = \frac{tp}{tp + fn} \tag{6}$$

pr and re are the measured precision and recall consecutively. tp is the number of true positives, fp is the number of false positives, and fn is the number of false negatives [15, 16]. Beside MAP, log average miss rate (LAMR) is used to evaluate detection performance [16].

$$\mathrm{LAMR} = \exp\left(\frac{1}{9} \sum_{f} \log\left(mr(\arg\max_{fppi(c) \leq f} fppi(c))\right)\right) \tag{7}$$

The nine $fppi$ reference points f are similarly spaced in the logarithmic space, such that $f \in \{10^{-2}, 10^{-1.75}, \ldots, 10^{0}\}$. Every $fppi$ that becomes reference point has the appropriate mr value to use. If the value of miss rate for a given f does not exist, then the maximum existing $fppi$ value is used as a new reference point, which is carried out by $mr(\arg\max_{fppi(c)\leq f} fppi(c))$ where mr and $fppi$ are defined as follows [16]:

$$mr(c) = \frac{fn(c)}{tp(c) + fn(c)} \tag{8}$$

$$fppi(c) = \frac{fp(c)}{\#img} \tag{9}$$

$tp(c)$, $fp(c)$ and $fn(c)$ are the number of true positives, false positives and false negatives consecutively. c is the confidence value where the detection results that are taken into account are those that have a confidence value equal to or greater than c. The confidence threshold c is used as a control variable.

3 Proposed Methodology

The population used in this study as training data and testing data are container images, while the trial sample in this study is video data that display containers. The process of collecting data is done by directly recording images or video of containers with the camera and also by crawling. The crawling process aims to harvest data sourced from the Internet based on container data searching. Software and application used are Anaconda, Python, Darfklow-YOLOv2, TensorFlow and LabelImg. YOLO interpretation method is used to detect and classify container objects. The proposed workflow of the system is shown in Fig. 2.

4 Results and Discussion

In the training process, some checkpoints are saved in the form of a file containing information about the weight and bias of the network to the extent that it has been trained based on certain steps. Five checkpoints are stored with random intervals. The following figure is a sample trial data of the results of container object detection in one image frame.

Based on Fig. 3, in the image frame, there are nine container objects, but which are stated to be detected are eight objects. Then the prediction of the detection results obtained on the container images that have been recorded in Table 1 is as follows:

The test results in Table 1 used the checkpoint number 2724. The results obtained from nine images of container, eight images were detected correctly as containers with confidence values ranging from the smallest 0.51 and the highest was 0.98. Figure 4 is the results of a model trial to detect containers in a video frame.

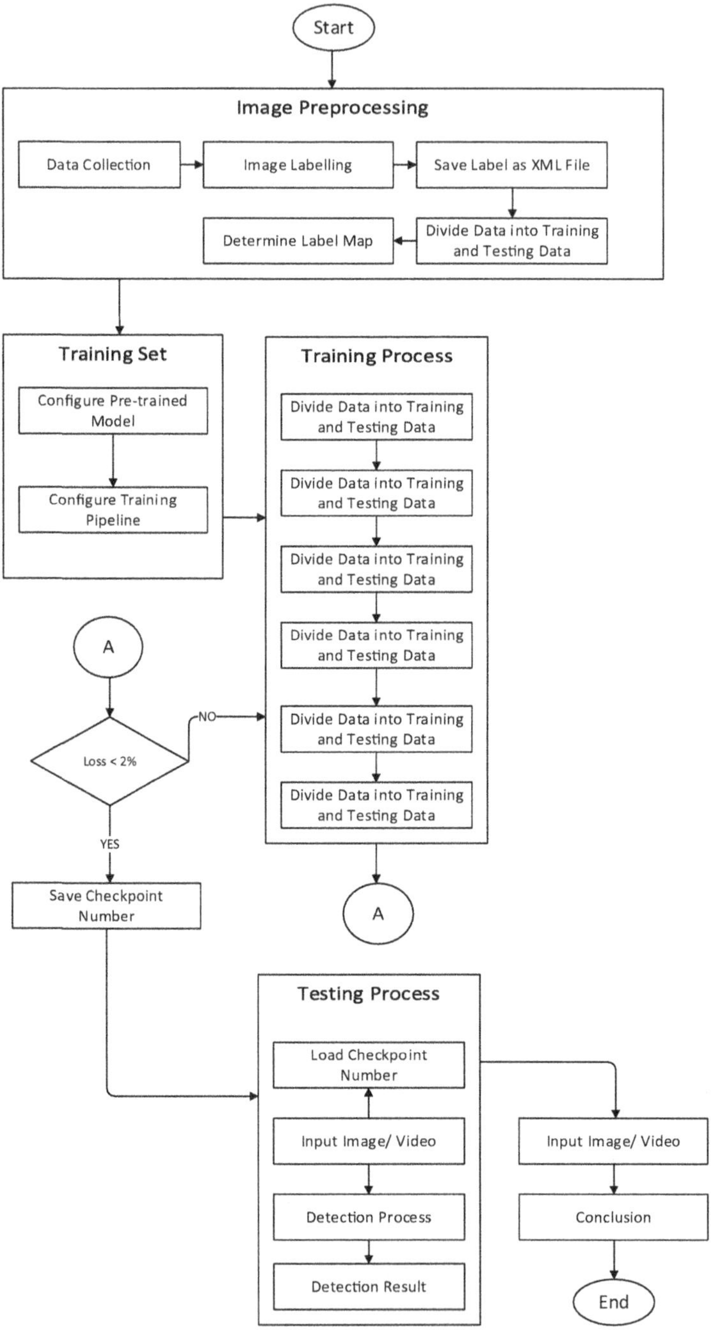

Fig. 2 Stages of research

Fig. 3 Sample of testing data

Table 1 Confidence level of container detection in Fig. 2

Label	Confidence	Top left	Bottomright
Container (1)	0.50567037	{'x': 3, 'y': 29}	{'x': 41, 'y': 80}
Container (2)	0.5136227	{'x': 72, 'y': 16}	{'x': 132, 'y': 73}
Container (3)	0.91867775	{'x': 47, 'y': 76}	{'x': 124, 'y': 164}
Container (4)	0.9010935	{'x': 125, 'y': 72}	{'x': 214, 'y': 164}
Container (5)	0.76766604	{'x': 0, 'y': 108}	{'x': 39, 'y': 172}
Container (6)	0.90811706	{'x': 1, 'y': 181}	{'x': 29, 'y': 250}
Container (7)	0.98076135	{'x': 51, 'y': 168}	{'x': 107, 'y': 250}
Container (8)	0.97507375	{'x': 131, 'y': 159}	{'x': 209, 'y': 255}

Based on Fig. 4, from two video frames, containers were marked with a square with confidence for the first image with a value of 0.9 and the second with a value of 0.3. To prove the confidence score based on the results of the prediction, the mean average precision (MAP) evaluation is implemented on several checkpoint files that have been tested. Then the results are presented in the table as follows:

From Table 2, checkpoint 2724 has the highest MAP that mean checkpoint 2727 is most precise in comparison with other checkpoint, while false positive and true positive are found as follows:

In accordance with Table 3, the evaluation results stated that the container objects from a total of 32 test data and container classes, for checkpoint number 2474 the results of true positive were 105 objects but the results that stated false positive

Fig. 4 Container detection in a video frame

Table 2 Checkpoint comparison table based on MAP

Checkpoint	MAP (%)
2474	68.12
2599	62.60
2724	68.63
2974	66.67
3099	63.96

Table 3 True positive and false positive comparison table

Checkpoint	True positive	False positive
2474	105	2
2599	96	1
2724	105	0
2974	102	0
3099	98	1

amounted to two container objects. Among the checkpoints, the best is checkpoint number 2724 with true positive results as many as 105 objects with results that state false positive is zero.

To find out the false positive rate per image (FPPI) for the overall test data based on several model checkpoints, it is proven by the log average miss rate with the following table:

The scale in Table 4 is the log average miss rate of all checkpoints. The highest average error was the number 2474 with the number 0.41, and the lowest average error was the number 2724 with the number 0.31. These values are obtained by returning the error detection result level on a comparison scale with GroundTruth.

Table 4 Log average miss rate comparison table

Checkpoint	LAMR
2474	0.41
2599	0.4
2724	0.31
2974	0.33
3099	0.38

5 Conclusion

In this research, object detection has been carried out to detect containers. Containers images were used as training data which were then tested on other container images data. Application of CNN method in the form of YOLO is used for container tracking in video data. The results of training process are saved into several checkpoints. The best checkpoint gives a quite accurate detection of containers.

Acknowledgements The first author is grateful to the Ministry of Education and Culture of the Universitas Pembangunan Nasional "Veteran" Jawa Timur, Faculty of Computer Science, Indonesia, who funded this research publication. E. Joelianto and P. Siregar are supported by the Ministry of Research, Technology and Higher Education under Higher Education Applied Research Grant 2018-2019, Indonesia.

References

1. Perkovic, M., Gucma, M., Luin, B., Gucma, L., Brcko, T.: Accommodating larger container vessels using an integrated laser system for approach and berthing. Microprocess. Microsyst. **2**, 106–116 (2017)
2. Yu, M., Qi, X.: Storage space allocation models for inbound containers in an automatic container terminal. Eur. J. Oper. Res. **226**(1), 32–45 (2013)
3. Henning, K., Bruns, M.: Design of an automatic control system for train-to-train container transfer. IFAC Proc. Vol. **16**(4), 165–173 (1983)
4. Wu, W., Liu, Z., Chen, M., Yang, X., He, X.: An automated vision system for container-code recognition. Expert Syst. Appl. **39**(3), 2842–2855 (2012)
5. Girshick, R, Donahue, J, Darrell, T, et al.: Rich feature hierarchies for accurate object detection and semantic segmentation. Comput. Sci, 580–587 (2013)
6. Girshick, R.: Fast R-CNN. IEEE International Conference on Computer Vision, 1440–1448 (2015)
7. Wei, L., Dragomir, A.: SSD: Single Shot MultiBox Detector. arXiv preprint arXiv:1512.02325v5 (2016)
8. Redmon, J., Divvala, S., Girshick, R., Farhadi, A.: You only look once: unified, real-time object detection. arXiv preprint arXiv:1506.02640 (2015)
9. Lan, W., Dang, J., Wang, Y., Wang, S.: Pedestrian detection based on YOLO network model. In: 2018 IEEE International Conference on Mechatronics and Automation (ICMA). doi:https://doi.org/10.1109/icma.2018.8484698 (2018)
10. Berg, A.: Learning to Analyze what is Beyond the Visible Spectrum (Vol. 2024). Linköping University Electronic Press (2019)

11. Goodfellow, I., Bengio, Y., Courville, A.: Deep learning. MIT Press (2016)
12. Rosebrock, A.: Deep learning for computer vision with python: starter bundle. PyImageSearch (2017)
13. Basha, S.H., Dubey, S.R., Pulabaigari, V., Mukherjee, S.: Impact of fully connected layers on performance of convolutional neural networks for image classification. arXiv preprint arXiv: 1902.02771 (2019)
14. Hendry, & Chen, R.-C.: Automatic license plate recognition via sliding-window Darknet-Yolo deep learning. Image Vision Comput. doi: https://doi.org/10.1016/j.imavis.2019.04.007 (2019).
15. Everingham, M., Van Gool, L., Williams, C.K., Winn, J., Zisserman, A.: The pascal visual object classes (voc) challenge. Int. J. Comput. Vision **88**(2), 303–338 (2010)
16. Braun, M., Krebs, S., Flohr, F., Gavrila, D.M.: The eurocity persons dataset: A novel benchmark for object detection. arXiv preprint arXiv:1805.07193 (2018)

Multi-Label Classification Using Problem Transformation Approach and Machine Learning on Text Mining for Multiple Event Detection

Hadi Safari and Kusprasapta Mutijarsa

Abstract Social media is a source that stores a lot of valuable information. One of which can be used to know the event that occurs in urban areas that are shared by the urban society. Information shared such as congestion and floods can be used as decision-making materials in city management. One of the social media that is popularly used is Twitter. A tweet sometimes contains more than one type of event information at the same time. That means the tweets are associated with more than one label or called multi-label classification. The purpose of this research is to find the best models for event detection from user's tweets into multi-label classification using problem transformation approach and machine learning (ML) techniques. In this paper, events detected are the natural events that were related to traffic and natural disasters. The results of multi-label classification experiments show the accuracy of 87.0%, F-score 89.1%, and hamming loss 9.2%. In this paper, we also calculated out of vocabulary (OOV), where the number of OOV tokens not found in the training data reached 79.96%. Nevertheless, the difference in vocabulary and OOV values is very small, the accuracy of 4.5%, F-score 5.2%, and hamming loss 3.2%.

Keywords Multi-label classification · Problem transformation · Machine learning · Event detection · Twitter classification · Text mining

1 Introduction

Multi-label classification is a specific form of classification, which is part of data mining task [1]. In contrast to the traditional classification that associates document exactly one label on the label set, multi-label classification is associated with multiple labels on label set [2]. There is two taxonomy used to solve multi-label problems are problem transformation methods and adaptation algorithm methods. In the problem transformation method, multi-label transformed into single-label classification and at

H. Safari (✉) · K. Mutijarsa
Institut Teknologi Bandung, Bandung, Indonesia
e-mail: hadisafari.hs@gmail.com

E. Joelianto et al. (eds.), *Cyber Physical, Computer and Automation System*, Advances in Intelligent Systems and Computing 1291,
https://doi.org/10.1007/978-981-33-4062-6_9

the end process is reunited, while the adaptation algorithm method is to use existing classification algorithm and manage in order to handle documents that have multiple label pairs.

Social media became one of the contributors to growth vast data. Vast amounts of data increase the needed for data processing automatically. Machine learning (ML) is an appropriate technique for that, especially for data classification tasks. Classification is a supervised learning in which computer program is given learning task based on the experience of the given example in order to label objects based on examples given previously. ML can be used to find relations between basic features and classes on objects. Therefore, ML is very compatible with the purpose of this research. Five selected algorithms are support vector machines (SVM), k-nearest neighbors (kNN), decision tree (DT), Naive Bayes (NB), and random forests (RF). These five algorithms have been successfully used for event detection and classification of text on social media [3–6].

Events detection using Twitter is not new tasks. Previously, this have been done to events detection such as earthquakes [3, 7], hurricanes [3, 8], traffic [4, 5, 9–11], and crimes [8, 12]. However, events detected are only categorized into the single-label classification. This is a research opportunity where the detected events of tweets can be categorized into multi-label classification.

In this paper, we propose event detection derived from user Twitter using problem transformation approaches and machine learning technique to traffic event detection and natural disasters into the multi-label classification. In addition, labeling is done automatically by the model in the prediction phase, which then becomes the training dataset for retraining.

The rest of paper is organized as follows: Sect. 2 summarizes previous research. Section 3 discusses the proposed methodology from the beginning of data collection to the end of the classification process. Section 4 describes source and amount of data used. Section 5 discusses phase and results experiment. Finally, Sect. 6 contains the conclusions result experiment.

2 Related Work

This section summarizes the related research divided into two groupings: (1) based on event type using Twitter data and (2) multi-label classification.

Based on event type, Sakaki et al. [3] investigated natural disasters event (earthquakes, hurricanes) in the real time and proposed monitoring system of events notification and sending notification on time. The classification type used is the binary classification. Nguyen et al. [7] proposed a method of convolution neural network (CNN) to detect earthquake events in real time, and tweets are classified into the binary classification. Salas et al. [4] investigated traffic events detection in real time based on geolocation, first tweets are classified into binary classification, then event-related tweets are classified into multi-class classification, and SVM is used as the classification algorithm. D'Andrea et al. [5] propose a real-time monitoring system to traffic event detection. Five algorithms are used for classification: SVM, NB, C4.5, kNN ($k = 1, 2, 5$), and PART. Datasets are classified into binary classification (traffic, non-traffic) and multi-class classification (traffic caused by external factors, traffic congestion or traffic, non-traffic). Pandhare et al. [9] investigated event detection of traffic and accident in real time. Logistic regression and SVM are used to classify tweets into the binary classification. Georgiou et al. [10] proposed a method to find the relation between traffic congestion that occurred with the react car commuters based on the volume of messages and complaints on Twitter. Tejaswin et al. [11] proposed a method to predict traffic accidents based on weather conditions and visibility levels, with RF classification algorithm. Data is grouped into the binary classification. Souza et al. [12] proposed methods for detecting crime information in real time. The information gain is used as a feature selection algorithm and uses three classification algorithms: SVM, NB, and DT.

Multi-label classification: Lee et al. [13] proposed an estimated bus arrival time method with four target labels and used problem transformation approach, adaptation algorithm, and ensemble method. ML algorithms used to the text classification are DT, RF, NB, neural networks model (ANN), and AdaBoost. In the field of genetics/biology, Fitriawan et al. [14] proposed a new way of filtering drugs into multi-target using a combination of deep belief networks (DBN) and binary relevance data transformation method. The number of target labels on the protease is 15 and the kinase is 26. Yamada et al. [15] proposed a social event detection system with a total of five target labels. Tweets are classified into binary classification (event, non-event) using SVM algorithm. The three ways to classification into multi-label are (1) using keywords for each category; (2) binary relevance problem transformation approach; (3) hybrid of the combination of keywords and ML.

3 Methodology

In general, the system built consists of three main processes that are collecting data tweets, preprocessing, and classification process. The proposed methodology consists of three stages as shown in Fig. 1. The proposed methodology can be used for binary classification, multi-class classification, and multi-label classification.

3.1 Data Collecting

Twitter provides free data access in two ways: rest API and streaming API. In this research, the rest API is used to obtain data used keyword parameters, while streaming API to obtain tweets in real time without keyword parameters. The data collected is tweets data from Bandung city society by crawling at latitude point -6.917464 and longitude 107.619123 and set the radius of 20 km. Rest API data is collected over a

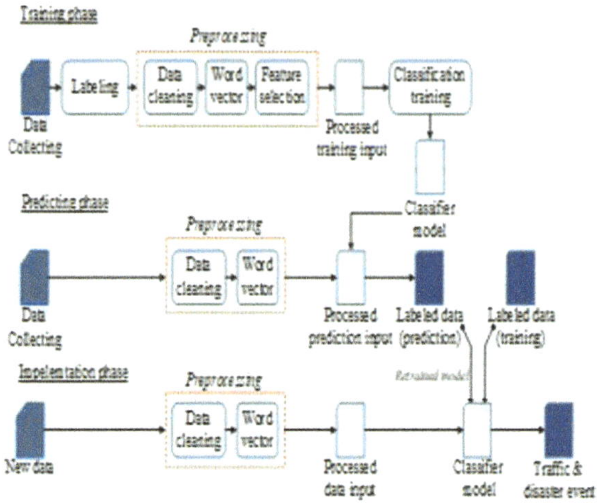

Fig. 1 System architecture for multi-label classification process

time span of November 16, 2017, to March 17, 2018, while streaming API data was collected on November 12, 2017, to November 13, 2017.

Traffic data collected refers to keywords K_1 = {"lalinbdg", "queue" (*macet*), "accident" (*kecelakaan*), "collision" (*tabrakan*)}, for natural disaster K_2 = {"flood" (*banjir*), "fire" (*kebakaran*), "avalanche" (*longsor*)"}. Data collection K_1 and K_2 is using OR operator, while multi-label data is a combination of keywords K_1 and K_2 using AND operator $K_3 = \{K_1 \cup K_2\}$.

3.2 *Labeling*

Annotator is manually labeling as training data input in training phase, while in prediction and implementation phase of data labeling using the model has been built. In the first labeling phase, the data is labeled into two categories which are event related and not. Furthermore, data event related is divided into traffic, disaster, and multi-label.

Datasets labeled are L_1 = {"event"}, L_2 = {"non-event"}, L_3 = {"traffic"}, L_4 = {"disaster"}. Data labeled on binary classification is a combination of L_3 and L_4 data labels, $L_1 = \{(L_3 \in L_1), (L_4 \in L_1)$ or $(L_3 \cup L_4)\}$. In multi-class classification, L_3 and L_4 are two separate data, while data on multi-label classification is tweets containing K_1 and K_2 or $L_5 = \{K_3\}$.

3.3 Preprocessing

This phase is preparing the data for the classification process consisting of three steps that are data cleaning, word vector, and feature selection. That is included in the preprocess category because it is to be done prior to the classification process.

1. Data cleaning: Data cleaning is using natural language processing (NLP) technique which consists of case folding, tokenization, URL removing, numbers removing, punctuation removing, and special characters except for question mark, expanding acronym, stemming, and stop words [16].
2. Word vector [17]: A document $D = \{d_1, d_2, \ldots, d_i\}$ represented as a dimension vector m, where each dimension is associated with each unique word $W = \{w_1, w_2, \ldots w_j\}$ and m is the total unique word in the document. $w_J = 1$ represents the words in document and $w_j = 0$ if the word is not in document.
3. Feature selection is a subset selection to find an optimal feature set that can represent documents from a number of great original features. Gini index (GI) [1] and Kruskal Wallis (KW) are feature selection algorithms with filter method. Both algorithms work by ranking all the features in the dataset or called feature ranking group.

3.4 Classification

This research used five ML algorithms as a comparison to get the best model of support vector machines (SVM), k-nearest neighbors (kNN), decision tree (DT), Naive Bayes (NB), and random forests (RF). The problem transformation approach becomes the way used to handle multi-label classification, it transforms multi-label data into the single-label, and two transformation methods used are binary relevance (BR) and label powerset (LP). BR method transforms multi-label into binary classification as shown in Fig. 2 and LP to the multi-class classification as shown Fig. 3.

Fig. 2 Multi-label transformation

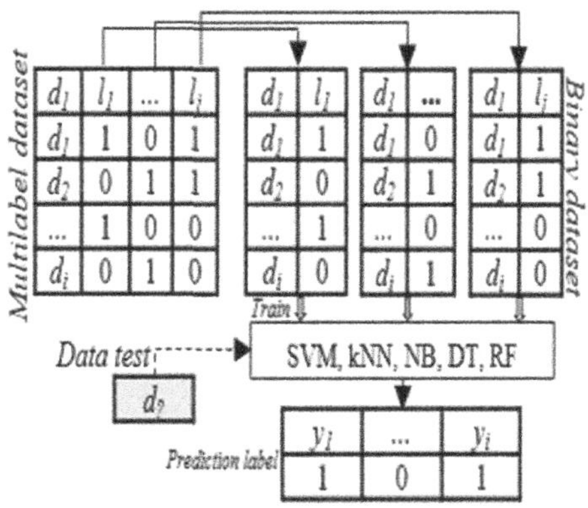

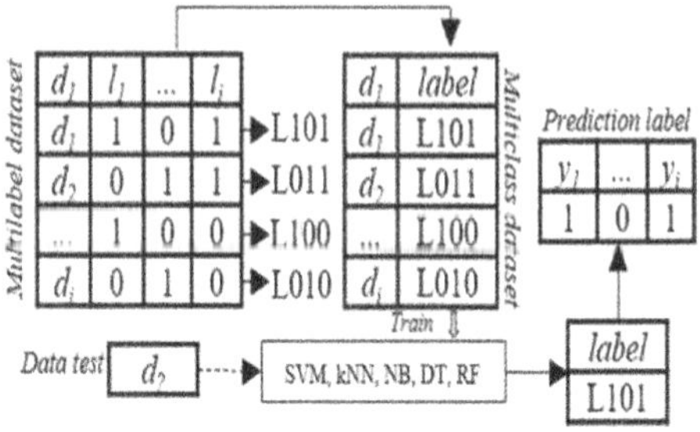

Fig. 3 Multi-label transformation using LP using BR

The label sets are binary classification $C_1 = \{L_1, L_2\}$, multi-class classification $C_2 = \{L_3, L_4, L_2\}$, and multi-label classification $C_3 = \{L_5, L_3, L_4, L_2\}$.

4 Data Analysis

The number of data collected is 33.823, with 28.070 through rest API and 5.753 through streaming API. Rest API is used to obtain data containing K_1 and K_2 keywords. Stream API is used to obtain data belonging to category L_2 but does not contain K_1 and K_2.

Distribution of data in each classification can be seen in Table 1. Dataset II is the addition of data from dataset I which is a combination of data labeled manually and data with labeling prediction results. For multi-label dataset, no data is added, but testing out of vocabulary (OOV).

The purposes of OOV testing to show the feature selection capabilities used are able to generate the built classification model. The amount of OOV testing data is 800.

Table 1 Spread of dataset

Classification	Label	Dataset I	Dataset II
Binary classification (A)	Event	375	1.714
	Non-event	375	1.714
	Traffic	250	857
Multi-class classification (B)	Disaster	250	857
	Non-event	250	857
	Multi-label	604	OOV (800)
Multi-label classification (C)	Traffic	604	OOV (800)
	Disaster	604	OOV (800)
	Non-event	604	OOV (800)

5 Experiment Results and Evaluation

Binary classification and multi-class classification testing are performed to determine whether the ML algorithm and feature selection that used just as well as on multi-label classification evaluation value.

5.1 Experiment Step

The main focus of testing is binary classification, multi-class classification, multi-label classification, feature selection model, and ML algorithm applied to all experiments. In all classification type, feature rankings are performed using k top ranking where k is 25, 50%, and all features of the selected dataset using GI and KW feature selection.

The step of an experiment is performed.

- Binary classification: After the labeling and preprocessing process, the dataset I (A) is trained using five machine learning algorithms with the tenfold cross-validation method.
- Multi-class classification: The stages performed the same as binary classification, but data used is the dataset I (B).
- The best model generated in the multi-class classification training process is used for labeling data at the prediction.
- The predicted data is then combined with the dataset I (B) into dataset II (B). Re-testing is done using dataset II (B) by the same procedure. The best model of this retraining process is used in the implementation phase.
- Because of $L_1 = (L_3 \cup L_4)$, then L_3 and L_4 of dataset II (B) are combined with L_1 in the dataset I (A) so as generate the dataset II (A). Retraining on binary classification is repeated with the same procedure using dataset II (A).
- Multi-label classification: The dataset I (C) was trained using a problem transformation approach using five ML algorithms applied to each of the BR and LP methods. The best model generated will be used on the OOV test.
- The OOV testing was performed using a new dataset of 800 tweets.

To evaluate classification performance generated on a single label measured [18] accuracy, precision, recall, and F1-measure, while in multi-label measured [19] accuracy, F1-measure, and hamming loss.

Python programming languages and natural language toolkit libraries (NLTK), sastrawi libraries are used in preprocess. For ML tasks, Weka is used in binary and multi-class classification, whereas Meka is used in multi-label classification.

5.2 Evaluation of Binary Classification

Figures 4, 5, 6 and 7 show the results of experiments performed on two different datasets at all *k* top feature rankings. SVM being the best algorithm outperforms all four other algorithms. The highest score is SVM on the top 50% ranked features.

Figures 4 and 5 show the SVM values obtained in the top 50% ranked features using GI feature selection: 86.6667% accuracy, 86.8% precision, 86.7% recall, and 86.7% F-score, while in the KW 85.3333% accuracy, 85.4% precision, 85.3% recall, and 85.3% F-score. Although the highest precision value in KW obtained RF is 85.7%, but on the other evaluation, SVM is better.

In the dataset II (A) as shown Figs. 6 and 7, SVM is still the best algorithm with the top 50% ranked features using GI feature selection: 86.2602% accuracy, 86.3% precision, 86.3% recall, and 86.3% F-score, while at KW accuracy 85.9102%, 85.9% precision, 85.9% recall, and 85.9% F-score. There are several algorithms in binary classification that have decreased evaluation value after the added data. At the top 25% ranked features using GI decline occurs in kNN, NB, RF, and KW algorithms occur in NB and RF. Top 50% ranked features using GI decrease in SVM, kNN, NB, RF, and KW algorithms that occur in kNN, NB, and RF. For all features, the decrease occurred in the NB in both feature selection GI and KW and RF which only decreased the precision value in feature selection KW.

The best combination of feature selection and ML algorithms is SVM, top 50% ranked features, GI feature selection, and dataset I (A).

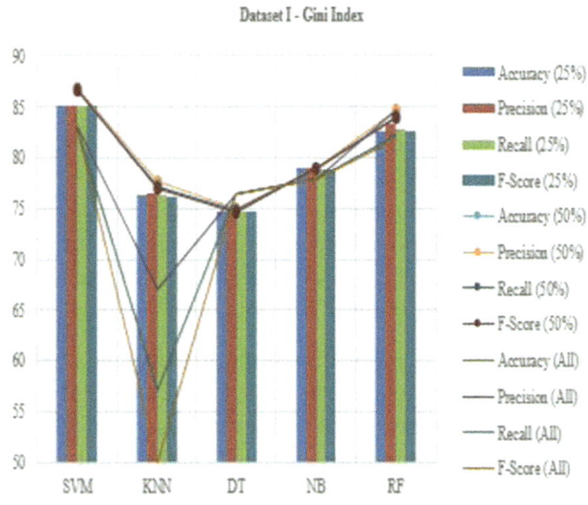

Fig. 4 Result comparison of binary classification using dataset I and Gini index feature selection

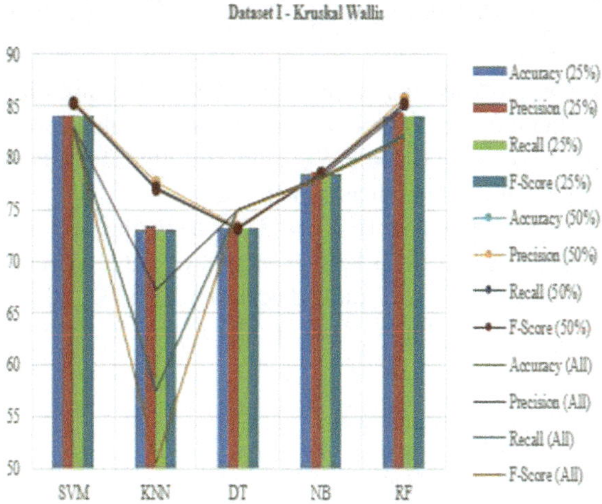

Fig. 5 Result comparison of binary classification using dataset I and Kruskal Wallis feature selection

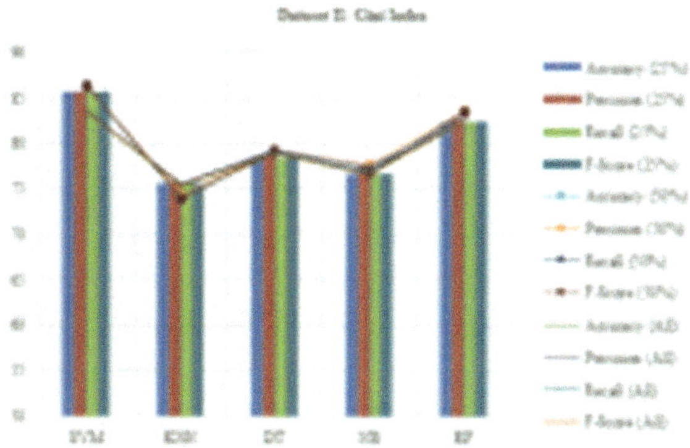

Fig. 6 Result comparison of binary classification using dataset II and Gini index feature selection

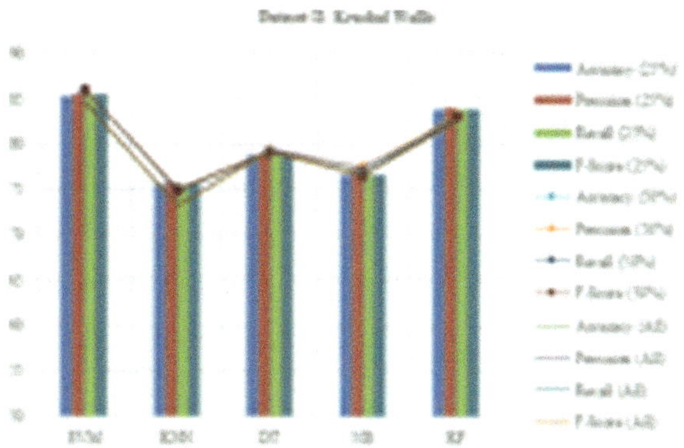

Fig. 7 Result comparison of binary classification using dataset II and Kruskal Wallis feature selection

5.3 Evaluation of Multi-Class Classification

Unlike previous experiments, SVM is the best algorithm in all top *k* ranked features, but in multi-class classification, RF is the highest evaluation value in top all features ranked of both GI and KW feature selection (Figs. 8 and 9), top 50% ranked features KW feature selection as shown Fig. 9.

Figures 8, 9 and 10 show SVM to be the best algorithm. Figure 8 on top 25% ranked features with 83.6% accuracy, 83.6% precision, 83.6% recall, and 83.5% F-score, while Fig. 7 ranked accuracy 83.3333, 83.3% precision, 83.3% recall, and 83.2% F-score. Figure 8 on top 50% ranked features with 85.8032% accuracy, 85.8%

Fig. 8 Result comparison of multi-class classification using dataset I and Gini index feature selection

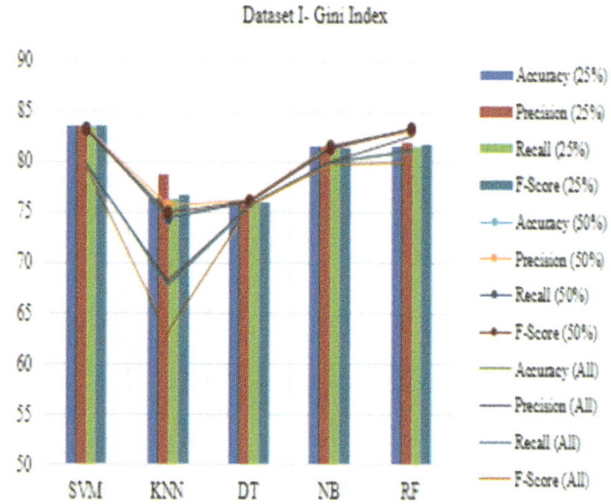

Fig. 9 Result comparison of multi-class classification using dataset I and Kruskal Wallis feature selection

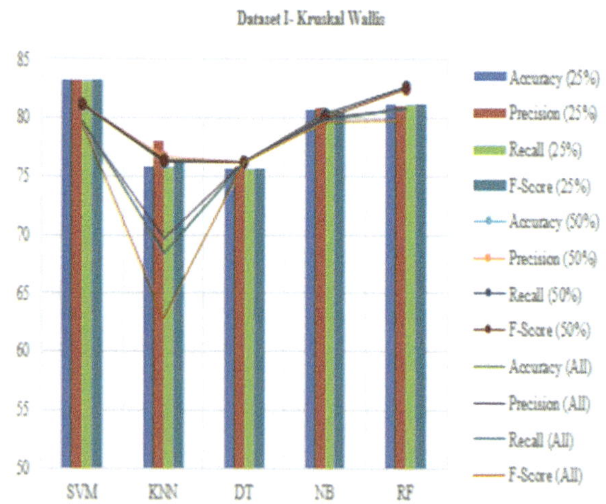

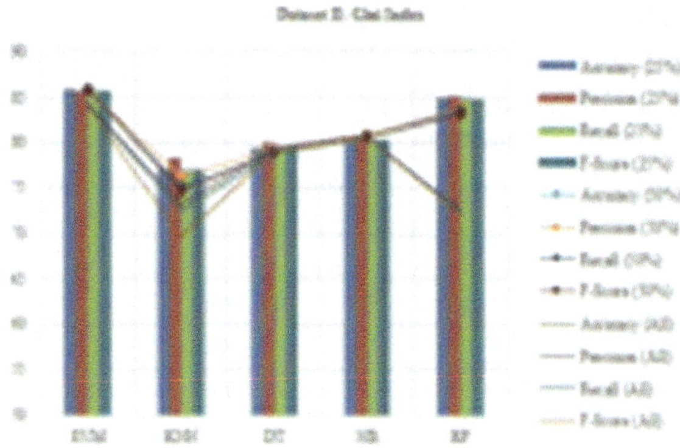

Fig. 10 Result comparison of multi-class classification using dataset II and Gini index feature selection

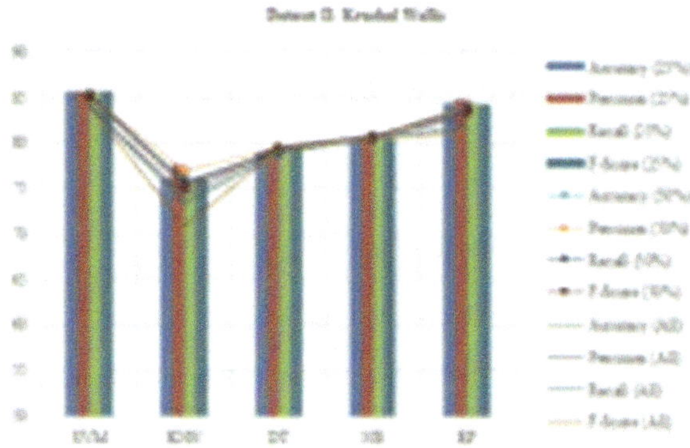

Fig. 11 Result comparison of multi-class classification using dataset II and Kruskal Wallis feature selection

precision, 85.8% recall, and 85.7% F-score. Figure 9 on top 25% ranked features with 85.6087% accuracy, 85.6% precision, 85.6% recall, and 85.6% F-score.

Impairment of evaluation also occurs in multi-class classification experiments after the addition of dataset. The NB algorithm has decreased evaluation value in the top 25% ranked features with GI and KW feature selection and top 50% ranked features in GI, while on all feature decrease occurs in RF with GI feature selection. The best combination of feature selection and ML algorithms is SVM, top 50% ranked features, GI feature selection, and dataset II (A) (Fig. 11).

5.4 Evaluation of Multi-Label Classification

In all of the top k ranked features, RF becomes the best algorithm, except for the LP method with the top 50% ranked features, SVM has the best value. Based on Tables 2, 3, and 4, overall LP is better than BR.

The best combinations of multi-label method, ML algorithm, and selection feature are LP, RF, all feature, and feature selection GI.

Table 2 Multi-label classification $K = 25\%$ top rank

				Feature 25%			
Multi-label method	ML algorithm		Gini index			Kruskal Wallis	
		Accuracy (%)	F-score (%)	H-loss (%)	Accuracy (%)	F-score (%)	H-loss (%)
	SVM	83.1	87.1	10.8	82.8	86.9	11.0
	KNN	80.5	84.3	13.1	79.8	83.6	13.7
BR	DT J48	79.5	83.8	13.5	84.9	88.1	9.9
	NB	73.1	79.4	18.3	74.9	80.6	17.1
	RF	84.7	87.8	10.2	84.9	88.1	9.9
	SVM	84.6	86.8	11.0	84.6	86.7	11.0
	KNN	81.1	84.2	13.1	79.9	83.1	14.0
LP	DT J48	80.9	84.2	13.2	81.5	84.6	12.8
	NB	77.9	81.3	15.7	79.4	82.7	14.5
	RF	84.9	87.5	10.5	85.5	87.8	10.2

Table 3 Multi-label classification $K = 50\%$ top rank

				All feature			
Multi-label method	ML algorithm	Gini index				Kruskal wallis	
		Accuracy (%)	F-score (%)	H-loss (%)	Accuracy (%)	F-score (%)	H-loss (%)
	SVM	83.8	87.6	10.4	84.5	88.2	9.9
	KNN	77.2	81.9	15.4	77.4	81.2	16.0
BR	DT J48	80.0	85.4	12.3	80.1	85.5	12.2
	NB	75.7	81.3	16.4	75.7	81.3	16.4
	RF	86.0	89.0	9.2	86.0	89.1	9.2
	SVM	86.1	88.2	9.8	86.6	88.7	9.4
	KNN	78.3	81.4	15.5	77.1	80.3	16.3
LP	DT J48	82.3	85.4	12.2	82.8	85.7	11.9
	NB	80.6	83.8	13.7	80.6	83.8	13.7
	RF	87.0	89.1	9.2	86.7	88.7	9.5

The OOV testing is performed using the best multi-label classification model of the test results as described in Sect. 5. Table 5 shows a comparison of training data and OOV testing data. Although the number of training data is more than OOV testing data, the number of OOV test data tokens is more. This shows tweets content used in OOV testing which is more diverse. The decline of value in vocabulary and OOV values is shown in Table 6.

Table 4 Multi-label classification K = all feature

				All feature			
Multi-label method	ML algorithm	Gini index				Kruskal wallis	
		Accuracy (%)	F-score (%)	H-loss (%)	Accuracy (%)	F-score (%)	H-loss (%)
	SVM	83.8	87.6	10.4	84.5	88.2	9.9
	KNN	77.2	81.9	15.4	77.4	81.2	16.0
BR	DT J48	80.0	85.4	12.3	80.1	85.5	12.2
	NB	75.7	81.3	16.4	75.7	81.3	16.4
	RF	86.0	89.0	9.2	86.0	89.1	9.2
	SVM	86.1	88.2	9.8	86.6	88.7	9.4
	KNN	78.3	81.4	15.5	77.1	80.3	16.3
LP	DT J48	82.3	85.4	12.2	82.8	85.7	11.9
	NB	80.6	83.8	13.7	80.6	83.8	13.7
	RF	87.0	89.1	9.2	86.7	88.7	9.5

Table 5 Evaluation of out of vocabulary (OOV)

Classification method	Dataset	Number of token	Number of token in vocabulary	Number of token OOV	OOV (%)
Multi-label classification	2.416	1.422	824	1.137	79.96
Testing OOV	800	1.961			

6 Conclusion

The proposed methodology proved is able to detect events categorized into the single-label and multi-label classification. The experimental results show that problem transformation approaches can be applied to traffic event detection and natural disasters derived from tweets that contain information of two events simultaneously. In general, the results obtained quite well with the accuracy rate reached 87, F-score 89.1%, and hamming loss 9.2%. LP performs better than BR on all combinations of ML algorithms and feature selection. SVM algorithm better than four algorithms on binary classification and multi-class classification. But in multi-label classification, RF becomes the algorithm that gives better performance than others. Overall, the five ML algorithms used resulted better performance when applied to multiple categories. Both were applied to the single-label and multi-label classification. This is indicated by the average evaluation results on multi-class classification which is better than binary classification. Similarly, when applied to multi-label classification, LP is better than BR.

Table 6 Comparison of measurement results between in vocabulary and OOV

In vocabulary					Out of vocabulary			Decline of value		
			Gini Index			Gini index				
Transformation method	ML algorithm	Accuracy (%)	F-score (%)	Hamming loss (%)	Accuracy (%)	F-Score (%)	Ham ming loss (%)	Accuracy (%)	F-score (%)	Hamming loss (%)
LP	RF	87.0	89.1	9.2	82.5	83.9	12.4	4.5	5.2	3.2

References

1. Gibaja, E., Ventura, S.: A tutorial on multilabel learning. ACM Comput. Surv. **47**(3), 1–38 (2015)
2. Sakaki, T., Okazaki, M., Matsuo, Y.: Earthquake shakes Twitter users: real-time event detection by social sensors. In: Proceedings of the 19th International Conference on World Wide Web (2010), pp. 851–860
3. Salas, A., Georgakis, P., Nwagboso, C., Ammari, A., Petalas, I.: Traffic event detection framework using social media (2017), p. 5
4. D'Andrea, E., Ducange, P., Lazzerini, B., Marcelloni, F.: Real-time detection of traffic from Twitter stream analysis. IEEE Trans. Intell. Transp. Syst. **16**(4), 2269–2283 (2015)
5. Xia, X., Togneri, R., Sohel, F., Huang, D.: Random forest classification based acoustic event detection (2017), pp. 163–168
6. Nguyen, V.Q., Yang, H.J., Kim, K., Oh, A.R.: Real-time earthquake detection using convolutional neural network and social data. In: 2017 IEEE 3rd International Conference on Multimedia Big Data (BigMM). IEEE (2017), pp. 154–157
7. Li, R., Lei, K.H., Khadiwala, R. and Chang, K.C.C.: TEDAS: a twitter-based event detection and analysis system. In: Proceedings of the International Conference on Data Engineering (2012), pp. 1273–1276
8. Pandhare, K.R., Shah, M.A.: Real time road traffic event detection using Twitter and spark. In 2017 International conference on inventive communication and computational technologies (ICICCT) (2017), pp. 445–449
9. Georgiou, T., Abbadi, A.E., Yan, X., George, J.: Mining complaints for traffic-jam estimation. In: Proceedings of the 2015 IEEE/ACM International Conference on Advances in Social Networks Analysis and Mining 2015 ASONAM (2015), pp. 330–335
10. Tejaswin, P., Kumar, R., Gupta, S.: Tweeting traffic : analyzing Twitter for generating real-time city traffic insights and predictions (2015)
11. Souza, A., Figueredo, M., Cacho, N., Araújo, D., Coelho, J., Prolo, C.A.: Social smart city: a platform to analyze social streams in smart city initiatives. In: Proceedings of the IEEE 2nd International Smart Cities Conference: Improving the Citizens Quality of Life, ISC2 (2016)
12. Kee, C.Y., Wong, L.P., Khader, A.T., Hassan, F.H.: Multi-label classification of estimated time of arrival with ensemble neural networks in bus transportation network (2017), pp. 150–154
13. Fitriawan, A., Wasito, I., Syafiandini, A.F., Amien, M., Yanuar, A.: Multi-label classification using deep belief networks for virtual screening of multi-target drug (2016), pp. 102–107
14. Yamada, W. et al.: Multi-label Categorizing Local Event Information from Micro-blogs. In: 2016 9th International Conference on Mobile Computing and Ubiquitous Networking (ICMU). IEEE (2016), pp. 1–6
15. Tala, F.: A Study of Stemming Effects on Information Retrieval in Bahasa Indonesia. M.Sc. Thesis, Append. D, pp. 39–46 (2003)
16. Salton, G., Wong, A., Yang, C.S.: A vector space model for automatic indexing. Commun. ACM **18**(11), 613–620 (1975)
17. Han, J., Kamber, M., Pei, J.: Data Mining: Concepts and Techniques, 3rd edn. Morgan Kaufmann, Waltham USA (2012)
18. Charte, F., del Jesus, M.J., Rivera, A.J.: Multilabel classification problem Analysis. In: Metrics and Techniques. Springer, Switzerland (2016)
19. Hardeniya, N.: NLTK Essentials: Build Cool NLP and Machine Learning Applications Using NLTK and Other Python libraries. Packt Publishing, Birmingham (2015)

Development of an Automatic Control System for Controlling of Soil pH Using a Microcontroller

Herriyance, Poltak Sihombing, and Rido Rivaldo

Abstract In agriculture, knowing the soil pH is very important not only to determine the right plants in the agriculture land but also to manage plant fertility. The main problem in this paper is how to monitor and control soil potential of hydrogen (pH) using a microcontroller and smartphone. Unfortunately, most of these pH detection systems still use conventional systems. This paper aims to describe the development of an automatic soil pH control system to become an intelligent farmer who can find suitable plants on agricultural land. The novelty of our approach is that sensors and microcontrollers can be used to find out the soil pH of farmland and are routed to smartphones automatically including its location using global positioning system (GPS). To support this research, we use the analog to digital conversion (ADC) signal method. We chose this model because it can be relied upon to detect soil pH precisely. Detection results can be used as a reference to decide also whether to reduce pH or increase soil pH. We have tested not only the type of soil pH in certain areas but also its location is included on Google maps.

Keywords Control system · Soil pH · Sensor · Microcontroller · Smartphone

1 Introduction

Soil is very vital for all life on earth because the soil supports plant life by providing nutrients and water to support plant roots. For a farmer, knowing how much of soil potential of hydrogen (pH) is the most important not only to determine what type of agriculture plant is suitable for the soil but also to manage the plant fertility.

The soil structure is not only good for roots but also for breathing for plants. Soil is also a habitat for various microorganisms, animals and life many people. Soil fertility can be known by the soil pH value. Measurement of soil acidity and soil pH value is also a parameter for determining soil fertility [1].

Herriyance · P. Sihombing (✉) · R. Rivaldo
Computer Science Department, Universitas Sumatera Utara, Medan, Indonesia
e-mail: poltak@usu.ac.id

E. Joelianto et al. (eds.), *Cyber Physical, Computer and Automation System*, Advances in Intelligent Systems and Computing 1291,
https://doi.org/10.1007/978-981-33-4062-6_10

The value of pH is the acidity level used to determine the acidity of the soil. The pH value is defined as the quantity logarithm of hydrogen (H^+) ion activity. The ion activity of hydrogen is difficult to measure experimentally, so the activity coefficient value is based on theoretical measurements. Therefore, the pH scale value is a relative value. This is a relative standard value as a solution to the pH value which is determined based on the international agreements [2].

The soil pH sometimes is called a pH meter. Unfortunately, many farmers do not have this tool may be because the price is quite expensive or lack of knowledge about the importance of knowing soil pH. Though the knowledge of the degree of soil acidity is very instrumental in the success of agriculture plants.

Plants will flourish and produce maximum if the soil pH is right for the plant. By knowing soil pH, farmers can determine the ideal pH which is suitable for plants. To get maximum results of a plant to be optimal, the soil pH must be in accordance with these plants. For example, corn can grow well at a place between 0 and 1300 m over sea level [3]. Corn plants will grow both on fertile soil, good drainage, warm temperatures of 21–32 °C, evenly distributed rainfall throughout the year, and monthly rainfall around 100–125 mm. Good soil for corn plants is soil with optimum pH 6.0–7.0.

To monitor the soil pH plants, we need a pH controlling tool that can detect automatically of soil pH. With the advances in computer science, especially on microcontroller, then revolution of controlling system based on Internet of things (IoT) is very possible to develop.

Besides soil pH, water vapor inside agriculture is one of the most significant variables affecting crop growth. High humidity may increase the probability of diseases and decrease transpiration. Low humidity may cause hydria stress, closing the stomata, and thus it may lower down the process of photosynthesis which depends on the CO_2 assimilation.

The humidity control is complex because if temperature changes then relative humidity changes inversely. Temperature and humidity are controlled by the same sensors. The main priority is for temperature control because it is the primary factor in crop growth. Based on the inside relative humidity value, the temperature set-point can be adjusted to control the humidity within a determined range.

Hence, to control the required soil pH and humidity is a very complex task. For proper control of humidity, internal air can be exchanged with outside air by properly controlling the ventilation of the agriculture area [4]. The main problem in this paper is how to monitor and control soil pH using a sensor, microcontroller and smartphone.

Michael Schirrmann and colleagues conducted a sensor readings study to evaluate pH on agricultural land in Germany [5]. Sensor reading is compared to the standard value of soil pH values. In [6], Kumar and colleagues described concepts and techniques for detecting soil pH. They said that soil pH is a very important parameter to increase crop productivity, so it must be handled appropriately.

In another study, Gaytri Gupta and colleagues developed an experimental tool to measure pH in certain types of substances [7]. This tool is used to check pH on substances found in daily life. According to their research, pH is a hydrogen potential that most important as a basis to measure the acidity of the soil. The pH value is to show the value of the negative logarithm of hydrogen ion concentration in some soil [8].

In [9], Sitompul et al. have developed the use of microprocessors for several fields of detection systems. Also, Sihombing et al. [10] have developed using a microcontroller for various purposes by connecting it to several sensors. Sihombing has developed a detection sensor whose results can be sent to the microcontroller and android smartphone [11]. Sihombing and colleagues have also developed a detection sensor that involves the location of an event and its results can be sent to the android smartphone in real time [12].

In other studies, Deepa Ramane et al. [13], wrote the integrated crop management systems have been designed to study spatial and temporal behavior of NPK and hydrogen ion concentration. Quantity of NPK and hydrogen ion concentration will determine the plant type and on plant growth fertility [14]. The soil pH environment condition will be the next benchmark for how much fertilizer will be added to the soil [15]. Also, in [16], Bachkar Yogesh Ramdas et al. have proposed that to increase crop productivity it needs continuous monitoring not only of soil pH but also moisture in automation in agricultural areas.

Based on the description above, it is important thing to know the pH of some soils for farmers or other users. For this reason, we need a pH meter model as described in this study.

This study proposes a model of pH sensor devices, to detect soil pH and plant references that are suitable for the position of the land. In addition to the soil pH, communication for sending of soil pH is also very necessary to find out the coordinates of the soil location correctly. It may need to involve sensor detection and communication media such as android for sending data in real time.

2 Method

This section describes the method of controlling a system for soil pH. We use the analog to digital conversion (ADC) signal method to support this research. We chose this model because it can be relied upon to detect soil pH precisely. Detection results can be used as a reference to decide also whether to reduce pH or increase soil pH.

The formula of an ADC method for calculating pH value is done by the following equation:

$$\mathrm{pH} = -\log\left[\mathrm{H}^{+}\right]\left(\text{neg. log of the } \mathrm{H}^{+} \text{ conc.}\right),$$

is reduced to $\mathrm{pH} = y$ with the following equation:

$$y = -6.159536842x + 24.74252632;$$

where y = pH value of nutrient, x = ADC value and Constanta (conc.) value = 24.74252632. With a value of 5 is the number of maximum voltage sensors, and 1023 is the maximum value of the ADC. The calculation example of y values as follows:

$$
\begin{aligned}
x_1 &= (\mathrm{ADC} * 5)/1023 \\
x_1 &= (594 * 5)/1023 \\
x_1 &= 2.903225806 \\
(x_1, y_1) &= (2.903225806,\ 6.86) \\
x_2 &= (\mathrm{ADC} * 5)/1023 \\
x_2 &= (689 * 5)/1023 \\
x_2 &= 3.367546432 \\
(x_2, y_2) &= (3.367546432, 4)
\end{aligned}
$$

The calculation of linear equations is as follows:

$$
\begin{aligned}
&(y - y_1)/(y_2 - y_1) = (x - x_1)/(x_2 - x_1) \\
&(y - 6.86)/(4 - 6.86) \\
&\quad = (x - 2.903225806)/(3.367546432 - 2.903225806) \\
&(y - 6.86)/(-2.86) = (x - 2.903225806)/0.464320626 \\
&0.464320626\,(y - 6.86) = -2.86(x - 2.903225806) \\
&0.464320626y - 3.185239492 = -2.86x + 8.303225806 \\
&0.464320626y = -2.86x + 8.303225806 + 3.185239492 \\
&0.464320626y = -2.86x + 11.4884653 \\
&y = -6.159536842x + 24.74252632.
\end{aligned}
$$

By the description above the value y = pH will be obtained by knowing the measure of ADC value resulted at that time. The pH probe detection resulted from the ADC will be converted into the soil pH automatically by the system. A simple diagram is shown in Fig. 1.

By using this system, a farmer can control a soil pH of agriculture plants. The brief description as follows:

A. *Simple Description of System*

The following is a brief description of how the system works:

- The user connects his smartphone with Arduino using Bluetooth by clicking on the button in the application.
- The user plugs the pH sensor of the probe into the ground where the pH value is calculated.
- The user clicks on the calculate button available in the application in order to calculate the pH value of the soil.

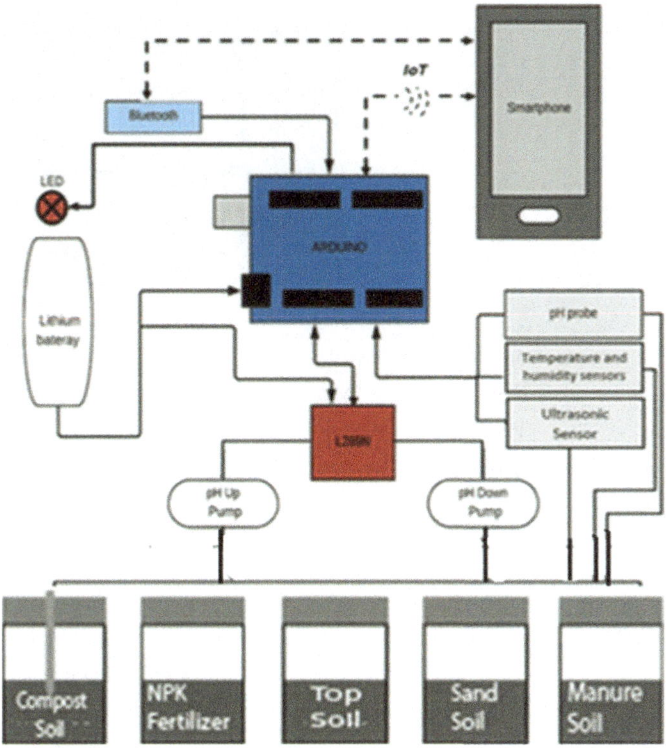

Fig. 1 Simple diagram of soil pH and humidity controlling

- Arduino will process and make calculations based on analog values received from the sensor.
- Arduino sends the soil pH value from the calculation results to the smartphone.

The user clicks on the Save button in the application to save the soil pH value on maps with the appropriate coordinates at the detection location.

B. *Soil pH and Humidity sensor*

- Soil pH sensor

We used the pH probe sensor in this research to detect acidity or alkalinity of soil. This sensor can measure pH with a very acidic concentration (pH 0) up to a very alkaline solution (pH 14) and can work at −10 up to 50 °C a water temperature. Figure 2 is shown the soil pH sensor, and sensor characteristics are shown in Table 1.

- Temperature and humidity sensor (HSM 20G).

This sensor is used to measure temperature and humidity in hydroponic plants as we shown in Fig. 3.

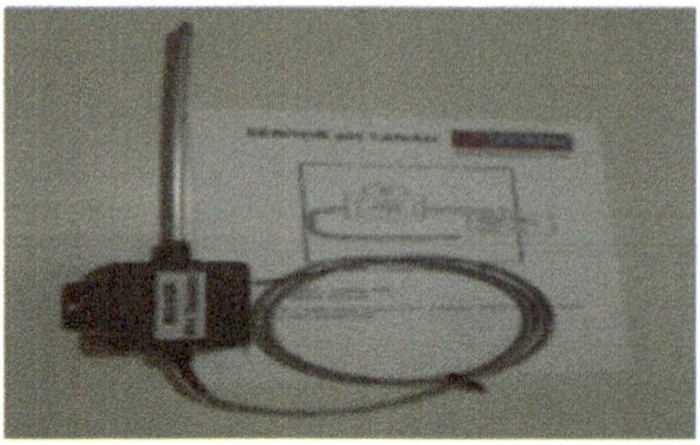

Fig. 2 Physical form of soil pH sensor

Fig. 3 Temperature and humidity sensor (HSM20G)

Table 1 Table captions should be placed above the tables

Parameter	Symbol	Min	Max	Units
Input voltage	*V*cc	3.0	4.7	V
Output voltage	ΔVolt	4	45	ADC
Response time	*T*	0.1	0.3	S
Sensitivity	*V*cc	0.036	0.234	V

- Degree of acidity

pH is the acidity or basicity of an object measured using a pH scale between 0 and 14. Acidic properties have a pH value between 0 and 7, and alkaline properties have a pH of 7–14 as we shown in Fig. 4.

3 Result and Discussion

A. *Software Arduino Uno*

To support this research, there are two stages of software implementation that we made, namely Arduino software and android software. We have developed the software for Arduino using C programming language and Arduino application as its compiler. The program file from the compiler has an.ino extension which is embedded

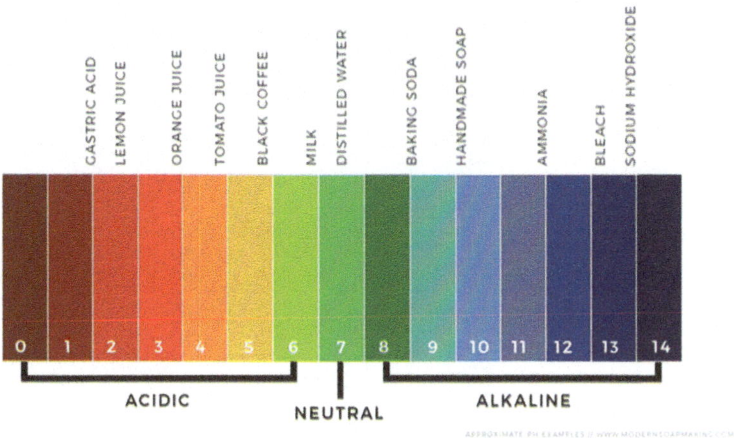

Fig. 4 Scale of acidity (pH)

in the Arduino microcontroller board. Figure 5 shows a part of developed software in Arduino microcontroller.

B. *Software developed in Android*

We have developed software application by Java programming in android smartphone that serves an information from Arduino microcontrollers. This software application contains an information of pH values from soil tested. To operate it, the user must connect the Arduino microcontroller through the smartphone in Internet network or Bluetooth. Figure 6 shows an example of experimental setup of soil pH application in android smartphone.

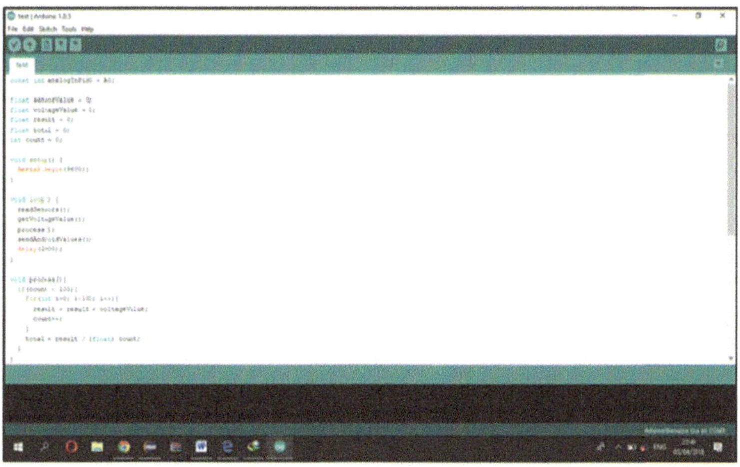

Fig. 5 Part of developed software of microcontroller

Fig. 6 Soil pH control in android application

The calculation results of soil pH have information in the form of what plants are suitable for planting according to the pH value that has been calculated. This display can be swiped, so users can see which plants are suitable for planting in this location. For example, Fig. 6 shows the results of the pH calculation and suitable plants to be planted.

With the ADC method, the sensor probe detection results will be directly converted into a soil pH value by the application system. An example of the results of this detection is shown in Fig. 5.

The application software developed will show not only the pH value of soil but also suggest the type of plants that are suitable to be planted at the detection location as shown in Fig. 5. In addition to displaying the pH value, the system also provides two buttons as an option to store the pH value and its location using GPS as shown in Fig. 7.

C. *The soil pH Android process resulted*

Testing the soil pH sensor is detected by using an analog signal as the transmission. This range of analog signals consists of 10 bits in range of 0–1023. Analog to digital

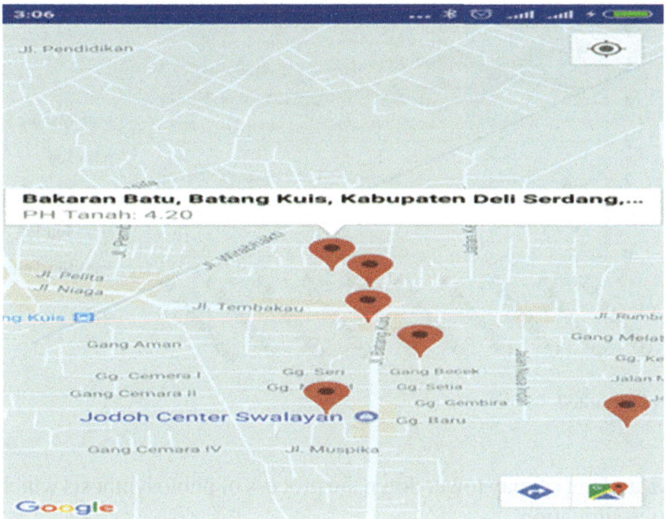

Fig. 7 Plants and locations based on GPS (in *Bahasa Indonesia*)

converter (ADC) signal be further processed by the Arduino and send it to a smartphone. The resulted of soil pH testing which is given a pH acid–base buffer solution based on a soil pH sensor datasheet is shown in Table 2.

Several experiments have been carried out with the detection results as shown in Fig. 8. These results indicate that the pH value measured by the soil pH sensor is not much different from the calculation of the soil pH manually. The voltage released by the sensor varies according to the pH level of some soil samples. The sensor takes about 2–5 s to calculate the pH value of the measured soil sample.

D. *Soil pH and Humidity*

Besides soil pH, water vapor inside agriculture is one of the most significant variables affecting crop growth. High humidity may increase the probability of diseases and decrease transpiration. Low humidity may cause hydria stress, closing

Table 2 Example of soil pH resulted

Soil sample	X	X_2	X_3	X_4	X_5	X_6	X_7	X_8	X_9	Avg
Manure soil	3.99	4.89	4.41	4.47	4.34	4.27	4.20	4.13	4.13	4.28
Compost soil	5.39	5.38	4.82	5.10	5.03	4.96	4.68	4.82	4.82	4.9
Sand soil	4.13	4.89	4.13	3.99	4.20	4.13	3.99	4.13	3.99	4.17
Humus	4.47	4.47	4.82	5.58	5.58	5.58	5.58	5.58	5.58	5.28
NPK fertilizer	1.77	2.12	1.84	1.77	1.98	2.12	2.26	1.98	2.19	2.01

where Avg = pH Average

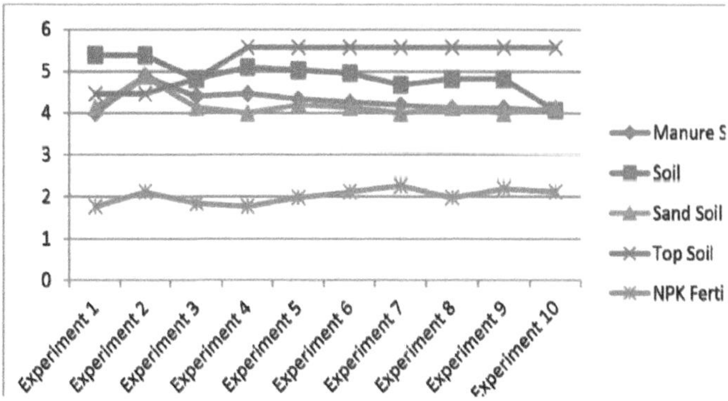

Fig. 8 Soil pH resulted

the stomata, and thus it may lower down the process of photosynthesis which depends on the CO_2 assimilation.

Temperature and humidity sensor is special devices consisting of characteristic semiconductor materials that change according to the amount of air humidity. Basically, this sensor is based on the number of water vapor. In this research, we use the HSM-20G sensor. This sensor can be converted to DC voltage which can be run with a digital system. Conversion between sensor output vs humidity is shown in Table 3.

To support this research, application software has been developed on an android smartphone that functions to control the soil pH of the plant. The part of apparatus experiment is shown in Fig. 9.

A. *Results of Soil pH and ADC*

By formula above:

$$y = -0.0693x + 7.3855$$

where

y = pH Value.

x = ADC Value.

We got the result as shown in Table 4.

The soil pH sensor uses analog signals as transmission. The range of this analog signal consists of 10 bits with a range of 0–1023. Table 4 shows the results of testing

Table 3 Soil pH resulted

Output (volt)	0.74	0.95	1.31	1.68	2.02	2.37	2.69	2.99	3.19
% RH (relative humidity)	10	20	30	40	50	60	70	80	90

Fig. 9 Soil pH controlling

Table 4 ADC and soil pH resulted

Soil pH	AVO meter (mV)	ADC	pH value by sensor
7	36	4	7.1083
7	41.5	6	6.9697
7	49.7	7	6.9004
6	117.9	20	5.9995
4.9	204	35	4.96
4.3	234	45	4.267

the soil pH sensor on soil that had previously been given an acid–base pH buffer solution.

4 Conclusion

This paper presented a monitoring and controlling soil pH, temperature humidity and location of plant in agriculture system. This system can be easily accessed through the browser of a smartphone. Microcontroller-based soil pH and temperature humidity measuring instruments have been developed in this study.

We have tested on several samples, such as soil fertilizer, compost, sand, humus and NPK fertilizer. The output of the measurement resulted as follows: The pH value of manure soil is 4.28; the pH value of compost soil is 4.9; the pH value of sand soil is 4.17; the pH value of humus is 5.28, and the pH value of NPK fertilizer is 2.01 as is shown in Table 2. In addition to pH measurement, the types of plants are adjusted to the measurements that notified. Also, the measurement location is notified by this application. The calculation results of the soil pH application are relatively not much different when compared to the calculation of manual soil pH. The voltage released by the sensor varies, according to the pH level of some soil samples. The sensor takes about 2–5 s to calculate the pH value of a soil sample. This paper also provides recommendations for suitable plants at the location of the land based on the pH value obtained.

Acknowledgements The authors would like to thank The Rector of Universitas Sumatera Utara (USU), the Chair of the Research Center of USU, and the Dean of Fac. of Computer Science and Information Technology, for The Research Award. This work was supported by "The Government of Republic Indonesia" under Research Grant 2019 in accordance with the Letter of Assignment Agreement, Number: ***198/UN5.2.3.1/PPM/KP-TALENTA USU/2019***, dated April 2019.

References

1. Fortier, J., et al.: Mature hybrid poplar riparian buffers along farm streams produce high yields in response to soil fertility assessed using three methods. Sustain. J. **5**(5), 1893–1916 (2013)
2. Cuilian, Y., et al.: Determination of pH value and acid-base property of ionic liquid aqueous solutions. Sioc. J. Acta Chimica Sinica **72**(4), 495–501 (2014). (Chinese Edition)
3. Liu, K., Wiatrak, P.: Corn plant characteristics and grain yield response to N fertilization prog. in no-tillage system. Am. J. Agric. Biol. Sci. **6**(1), 172–179 (2011)
4. Barbosa, G.L., et al.: Comparison of land, water, and energy requirements of lettuce grown using hydroponic versus conventional agricultural methods. Int. J. Environ. Res. Public Health. **12**(6), 6879–6891 (2015)
5. Schirrmann, M., et al.: Soil pH mapping with an on-the-go sensor. Int. J. Sens. **11**(1), 573–598 (2011)
6. Kumar, S., et al.: Soil pH sensing tech. and technologies-A review. IJAREEIE J. **4**(5), 4452–4456 (2015)
7. Brouder, S., Hofmann, B.S., Morris, D.: Mapping soil pH: accuracy of common soil sampling strategies and estimation techniques. Soil Sci. Soc. Am. J. **69**(2), 427–442 (2005)
8. Hashim, U., Haron, M.: Design of digital display for ISFET pH sensor by using PIC microcontroller. MASAUM J. Basic Appl. Sci. **1**(2), 326–330 (2009)
9. Sitompul, D., Sihombing, P.: A teaching media of using the busy bit and SDCC in display. Char. string on LCD in MCU 8051 IDE. Alexandria Eng. J. **57**(2), 813–818 (2018)
10. Sihombing, P., Karina, N.A., Tarigan, J.T., Syarif, M.: Automated hydroponics nutrition plants syst. using arduino based on android. J. Phys. Conf. Ser. **978**(1), 012014 (2018). doi: https://doi.org/10.1088/1742-6596/978/1/012014
11. Sihombing, P., Astuti, T.P., Herriyance, Sitompul. D.: Microcontroller based automatic temp. Control for oyster mushroom plants. J. Phys. Conf. Ser. **978**(1), 012031 (2018). doi: https://doi.org/10.1088/1742-6596/978/1/012031
12. Sihombing, P., Siregar, Y., Tarigan, J.T., Jaya, I., Turnip. A.: Development of building security integration syst. using sensors, microcontroller and gps (global positioning system) based android smartphone. J. Phys. Conf. Ser. **978**, 012105 (2018). doi: https://doi.org/10.1088/1742-6596/978/1/012105

13. Deepa, V. Ramane, et al.: Detection of NPK nutrients of soil using fiber optic sensor. Int. J. Res. Advent Technol. (E-ISSN: 2321–9637). Special Issue National Conference. 66–70 (2015)
14. Kulkarni, Y., Warhade, K.K., Bahekar, S.: Primary nutrients determination in the soil using UV spectroscopy. Int. J. Emerg. Eng. Res. Technol. **2**(2), 198–204 (2014)
15. Odum, H.T.: Environment, Power, and Society for the Twenty-First Century: The Hierarchy of Energy. Columbia University Press, USA (2007)
16. Ramdas, B.Y., Galande, S.G.: Green growth management by using arm controller. Int. J. Eng. Res. Appl. **4**(3), 360–363 (2014)

Influences of Off-Ramp Volumes on Mean Speed Based on METANET Model

Zahrotul Amalia Solakha, Salmah, and Imam Solekhudin

Abstract As a part of traffic control, METANET model is developed based on real situation related to influences of the existence of off-ramp volumes. Development of the model is based on models that have been found before, which is later derived so that we get the speed drop factor in discrete form. The effect of the off-ramp volumes is discussed in numerical implementation with some different traffic demand scenarios.

Keywords Traffic control · METANET · Mathematical model · Off-ramps effect · Mean speed

1 Introduction

In this modern era, transportation system is an important instrument which is needed by every people to support their necessity or activity. Along with inhabitant growth and development in many sectors, transportation system causes many problems such as traffic accident and traffic jam. Moreover, there are some problems that appear, especially from the traffic jam, such as time and energy losses, poor environmental quality, economics, social, and also health problems [1]. An effort that people can do to decrease traffic jam is traffic controlling. Traffic controlling needs mathematical model of the traffic problem.

Lighthill and Whitham in [2], Richards in [3], and Payne in [4] did some studies about traffic flow models. The model continuously developed based on the previous

Z. A. Solakha (✉) · Salmah · I. Solekhudin
Department of Mathematics, Universitas Gadjah Mada, Yogyakarta, Indonesia
e-mail: za.solakha@gmail.com

Salmah
e-mail: syalmah@yahoo.com

I. Solekhudin
e-mail: imams@ugm.ac.id

E. Joelianto et al. (eds.), *Cyber Physical, Computer and Automation System*, Advances in Intelligent Systems and Computing 1291,
https://doi.org/10.1007/978-981-33-4062-6_11

researches. Papageorgiou et al. in [5] observed nonlinear macroscopic model of traffic flow in Boulevard Peripherique, Paris. In 1990, Messmer and Papageorgiou introduced METANET model as a macroscopic simulation program for freeway network [6]. This model is a nonlinear discrete model. The METANET model also continuously developed based on the previous researches. In [7], an extension of the single-class macroscopic traffic flow model was proposed. METANET model is a model that can be used to control the traffic. This model focused on traffic situation in every street segment as a part of the link. There are some variables that used in this model: mean speed, traffic density, traffic flow, queue length, and outflow from an origin. (see [5, 8]).

To derive the model, there are some points to consider as a step to obtain the most suitable model based on the real situation. Some of the important aspects to consider are some particular cases in the traffic network. In this METANET model, the example of certain cases is the traffic network that contains on-ramps or street constriction. These cases result in a decrease of speed in the related street segment, just like the subsidiary model from Papageorgiou et al. in [9]. In [10], Hegyi declared that model of speed reduction factor caused by those two certain cases is in discrete forms toward segment i, link m, and time step k. For traffic network case related to off-ramps, Papageorgiou et al. in [5] wrote a continuous model from the speed reduction factor resulted by the case. Meanwhile, Hegyi gave a discrete equation from mean speed in the segment related to the off-ramps (see [10]).

It is necessary to make a flow traffic model which is suitable to the reality and ready to be implemented. In [11, 12], some strategies are proposed that aim to prevent the off-ramp queue spill-over and the mainstream congestion. In [13], Spiliopoulou et al. presented a real-time merging traffic control algorithm to reduce the freeway congestion due to an over-spilling off-ramp. In this paper, we discuss effect off-ramp volumes on mean speed on a freeway segment with some traffic demand scenarios. This model is then implemented numerically to solve problem involving link with off-ramp. One of the demand scenarios is taken from [14].

2 METANET Model

We refer to [6, 8], and [10] for a detailed description of the METANET. Below is basic explanation of METANET model.

Given link m that consists of N_m segments. In time step k, segment i, link m, these equations hold:

$$q_{m,i}(k) = \rho_{m,i}(k) v_{m,i}(k) \lambda_m \tag{1}$$

$$\rho_{m,i}(k+1) = \rho_{m,i}(k) + \frac{T_s}{L_m \lambda_m} \big(q_{m,i-1}(k) - q_{m,i}(k)\big) \tag{2}$$

$$\begin{aligned} v_{m,i}(k+1) = v_{m,i}(k) + \frac{T_s}{\tau}\big(V(\rho_{m,i}(k)) - v_{m,i}(k)\big) \\ + \frac{T_s}{L_m} v_{m,i}(k)\big(v_{m,i-1}(k) - v_{m,i}(k)\big) \\ - \frac{\eta T_s}{\tau L_m}\left(\frac{\rho_{m,i+1}(k) - \rho_{m,i}(k)}{\rho_{m,i}(k) + \kappa}\right) \end{aligned} \quad (3)$$

$$V\big(\rho_{m,i}(k)\big) = \min\left\{ v_{\text{free}} \exp\left[-\frac{1}{a_m}\left(\frac{\rho_{m,i}(k)}{\rho_{cr,m}}\right)^{a_m}\right], (1+\alpha)v_{\text{control},m,i}(k) \right\} \quad (4)$$

where $q_{m,i}(k)$, $\rho_{m,i}(k)$, $V\big(\rho_{m,i}(k)\big)$, and $v_{m,i}(k)$ is traffic flow (vehicles/hour), traffic density (vehicles/hour/lane), desired speed (km/hour), and mean speed (km/hour) in segment i, link m, at time step k, respectively. The notation λ_m stated number of lane, L_m stated length of street segment (m), and T_s stated duration of every time step. Symbol τ, η, κ, and a_m are model parameters. In the Eq. (4), $v_{\text{free},m}$ is vehicle's speed when the traffic is freely flowing in link m, $\rho_{cr,m}$ is critical density, and $v_{\text{control},\,m,i}(k)$ is speed limit permitted in segment i, link m, and time step k. The value of $(1+\alpha)$ stated that there is a probability of vehicles that breaks the fixed speed limit.

In time step k and origin o, these equations hold:

$$w_o(k+1) = w_o(k) + T_s[d_o(k) - q_o(k)] \quad (5)$$

$$q_o(k) = \min\left\{ d_o(k) + \frac{w_o(k)}{T_s}, r_o(k)C_o, C_o\left(\frac{\rho_{jam,m} - \rho_{m,1}(k)}{\rho_{jam,m} - \rho_{cr,m}}\right)\right\} \quad (6)$$

where $w_o(k)$, $q_o(k)$, $d_o(k)$, and $r_o(k)$ are queue length (vehicles), outflow (vehicles/hour), traffic demand, and ramp-metering rate in origin o, respectively. In Eq. (6), $\rho_{jam,m}$ is the maximum density in link m, and C_o is the capacity of origin o (vehicles/hour).

For the link that on-ramp is connected, speed drop factor is given by

$$-\frac{\delta_{\text{on}} T_s q_o(k) v_{m,1}(k)}{L_m \lambda_m \big(\rho_{m,1}(k) + \kappa\big)} \quad (7)$$

is required to add into the right-hand side of Eq. (3), where δ_{on} as a parameter. When the amount of lanes is decreased, speed drop factor is given by

$$-\frac{\emptyset T_s \Delta\lambda_m \rho_{m,N_m}(k) v^2_{m,N_m}(k)}{L_m \lambda_m \rho_{cr,m}} \quad (8)$$

is required to add into the right-hand side of Eq. (3) where Ø as parameter and $\Delta\lambda_m$ as the amount of lane decrease.

3 Influences of Off-Ramp Volumes on Mean Speed

In general, vehicles which turn out from main street decrease its speed so that they can enter the desired street safely. The speed of vehicles before they turn out from main street through off ramp may vary, depend on angle between the main street and the off-ramp. The decrease in the vehicles speed results in lower value of mean speed in the main street.

Based on [5], Eq. (3), which is the equation of mean speed, is obtained from the derivation of traffic flow model from [4] in [5], as

$$v(x, t + \tau) = V[\rho(x + \Delta x, t)]. \tag{9}$$

Following [5], the following equation can be obtained from Eq. (9).

$$\frac{\partial v}{\partial t} = -v\frac{\partial v}{\partial x} + \frac{1}{\tau}\left[V(\rho) - v - \frac{\eta \partial \rho}{\rho \partial x}\right] \tag{10}$$

where τ is time parameter. Equation (10) that has been transformed into discrete form toward link m, segment i, and time step k is equivalent to Eq. (3).

We assume that on link m, there is an off-ramp as illustrated in Fig. 1. Suppose that γ is an angle between main street and the off-ramp. Certain values of γ result in decrease in mean speed $v_{m,N_m}(k+1)$ in last segment of link m. By the existence of off-ramp which is connected to the link, speed of vehicles that turn out will be slower. This case influences the value of mean speed in street segment that is connected to the off-ramp. Let m be the index of the link which contains an off-ramp in the last segment, while $m+1$ is the index of the link which is located after the freeway off-ramp. We will find out how much the reduction of mean speed in the last segment of link m.

Let v be mean speed of link m. In general, vehicles that will turn out from the last link m have speed v_a. The value of v_a is not greater than mean speed in link m, adapted with angle γ. In the Eq. (3), assume that traffic flow which will turn out from point x by exiting rate $s(x, t)$ (vehicles/(hour.km)) in the speed v_a. The traffic flow proportion which will turn out through the off-ramp μ in the time step k is $\beta_\mu(k)$. In general, $v_a < v$ holds so that the value change $\partial v/\partial t$ is influenced by additional factor, that is, $\frac{s}{\rho}(v_a - v)$. The speed when vehicles are about to turn out from the street can be denoted as $v_a = \delta'_{\text{off}} v$, for $\delta'_{of f} \in [0, 1]$. Consequently, that additional

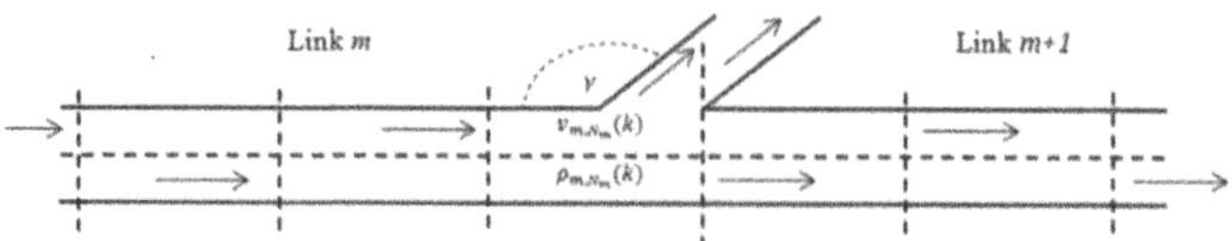

Fig. 1 When there is an off-ramp connected to the freeway the speed $v_{m,i}(k+1)$ is reduced

factor becomes

$$\frac{s}{\rho}(v_a - v) = \frac{s}{\rho}\left(\delta'_{\text{off}}v - v\right) = -\frac{\left(1 - \delta'_{\text{off}}\right)vs}{\rho} = -\frac{\delta_{\text{off}}vs}{\rho}, \tag{11}$$

where $\delta_{\text{off}} \in [0, 1]$. Since the change of $\partial v/\partial t$ is influenced by additional factor in Eq. (11), then, in this case, $-\frac{\delta_{\text{off}}vs}{\rho}$ needs to be added into the right-hand side of Eq. (10), therefore, we obtain

$$\frac{\partial v}{\partial t} = -v\frac{\partial v}{\partial x} + \frac{1}{\tau}\left[V(\rho) - v - \frac{\eta\partial\rho}{\rho\partial x}\right] - \frac{\delta_{\text{off}}vs}{\rho}. \tag{12}$$

Next, Eq. (12) is transformed into discrete form, similar to derivation process of Eq. (3) with $s(k) = \frac{\beta_\mu(k)q_{m,N_m}(k)}{L_m\lambda_m}$ yield

$$\begin{aligned} v_{m,i}(k+1) = v_{m,i}(k) &+ \frac{T_s}{\tau}\left(V(\rho_{m,i}(k)) - v_{m,i}(k)\right) \\ &+ \frac{T_s}{L_m}v_{m,i}(k)\left(v_{m,i-1}(k) - v_{m,i}(k)\right) \\ &- \frac{\eta T_s}{\tau L_m}\left(\frac{\rho_{m,i+1}(k) - \rho_{m,i}(k)}{\rho_{m,i}(k) + \kappa}\right) \\ &- \frac{\delta_{\text{off}}T_s\beta_\mu(k)q_{m,N_m}(k)v_{m,N_m}(k)}{L_m\lambda_m\left(\rho_{m,N_m}(k) + \kappa\right)}. \end{aligned} \tag{13}$$

Hence, the speed drop factor in the last segment of link m as the result of the existence of freeway off-ramp is

$$-\frac{\delta_{\text{off}}T_s\beta_\mu(k)q_{m,N_m}(k)v_{m,N_m}(k)}{L_m\lambda_m\left(\rho_{m,N_m}(k) + \kappa\right)}. \tag{14}$$

Value of vehicle's speed reduction due to off-ramp in time step $k + 1$ is directly proportional to a simulation time step T_s, parameter δ_{off}, proportion of traffic flow out from the main street, traffic flow and mean speed, and inversely proportional to traffic density in time step k, length of segment and number of lane on segment connected to the off-ramp. The higher of proportion of traffic flow out from the main street to the off-ramp, the larger decrease in vehicle's speed in the segment. Value of δ_{off} depends on proportion of the average of vehicle's speed turn out the main street. The smaller proportion, the higher value of δ_{off}. For instance, if the proportion is 0.6 times of average speed of vehicle on the main street, then $\delta_{\text{off}} = 1 - 0.6 = 0.4$. Since the average speed of the vehicles turning out the main street depends on angle γ, value of δ_{off} also depends on the value of γ.

4 Numerical Implementation

In this section, numerical implementation is executed in the network connected to off-ramp and the network that just consists of the straight street. The aim of the implementation is to compare mean speed, so in this case, it is not yet apply traffic control. Both of the network consists of six segments and two lanes. Length of each segment is 1 km and the origin is at the beginning of the first segment. These networks are illustrated in Fig. 2. In addition, these networks will be simulated in MATLAB® with different traffic demand scenarios to evaluate the off-ramp effect.

We use standard parameter $v_{\text{free},m} = 102$ km/h, $T_s = 10$ s, $\tau = 18$ s, $a_m = 1.867$, $\kappa = 40$ vehicles/km/lane, $\eta = 60\,\text{km}^2/\text{h}$, $\rho_{jam,m} = 180$ vehicles/km/lane, $\delta_{\text{on}} = 0.0122$, $C_o = 4000$ vehicles/hour, $\rho_{cr,m} = 33.5$ vehicles/km/lane, $L_m = 1$ km, and $\alpha = 0.1$, as used by [15, 16]. For parameter δ_{off}, we choose $\delta_{\text{off}} = 0.4$.

Numerical implementation is done by simulating METANET model in the both of the network for 60 min. The initial condition of street consists of initial mean speed and initial density in every segments and also the length of origin o initial queue. The traffic demand scenarios are assumed to be non-constant for 60 min (see Fig. 3). The second demand scenario is taken from [14], but it is implemented just for 1 h. The other assumption is the proportion of traffic flow of vehicles that are turned out from the main street for 60 min is depicted in Fig. 4.

Figures 5, 7, and 9 are the result of the numerical implementation on straight street without off-ramp, while Figs. 6, 8 and 10 are the result of the numerical implementation on straight street with off-ramp. Mean speed and traffic density in each segment, from first segment until last segment, are illustrated by red, magenta, cyan, blue, green, and black curve, respectively. For simulation results with traffic demand scenario 1, 2, and 3, there are speed reduction on the segment which is connected to off-ramp (blue curve).

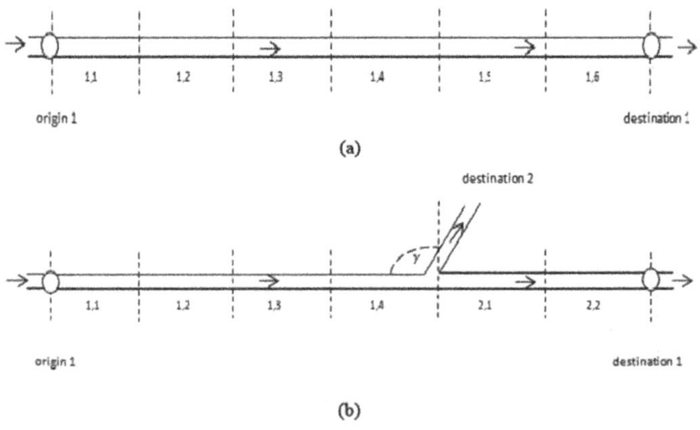

Fig. 2 Two traffic networks on this numerical implementation: **a** network 1: straight street consist of six segments without off-ramp and **b** network 2: six segments with off-ramp

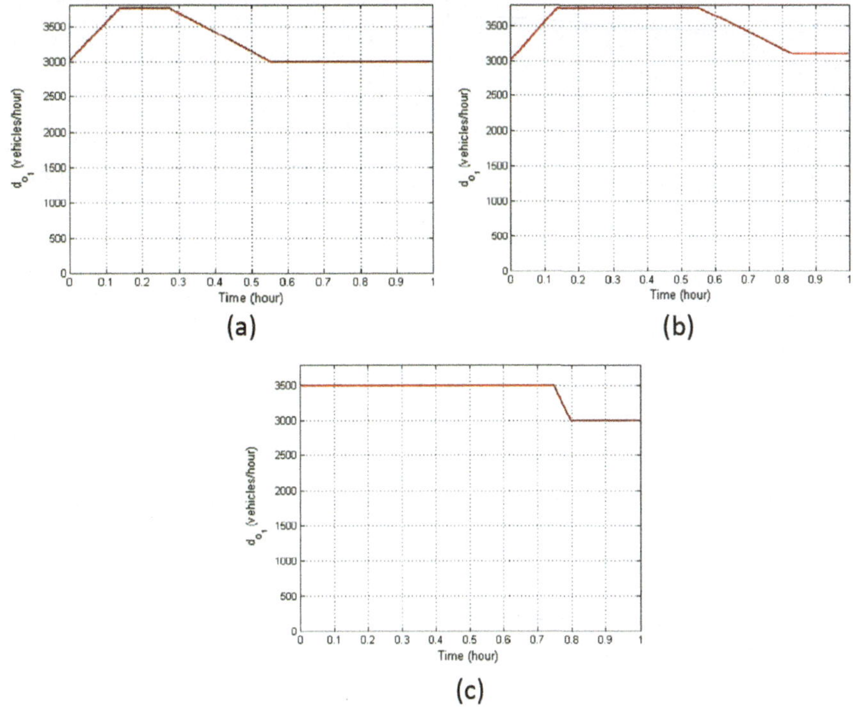

Fig. 3 Demand scenario in the simulation: **a** scenario 1, **b** scenario 2, **c** scenario 3

Fig. 4 Proportion of traffic flow of vehicles that are turned out from the main street for 60 min

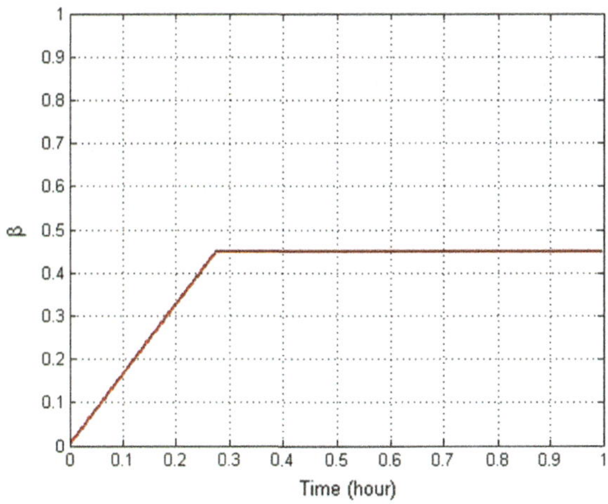

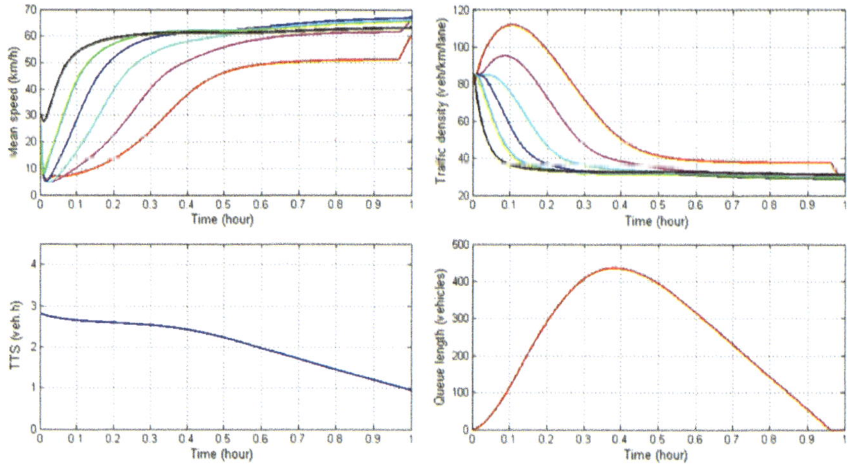

Fig. 5 Simulation result for the first traffic network (without off-ramp) with demand scenario 1: mean speed (km/hour), traffic density (veh/km/lane), queue length (veh), and TTS (veh.hour)

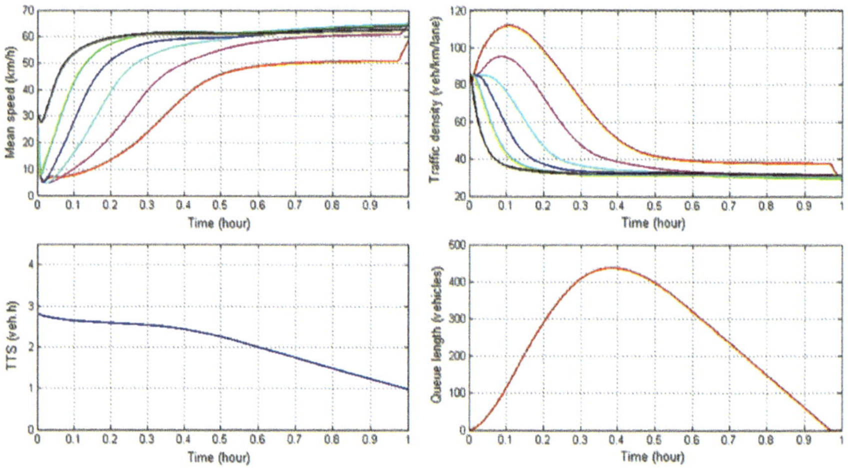

Fig. 6 Simulation result for the second traffic network (with off-ramp) with demand scenario 1: mean speed (km/hour), traffic density (veh/km/lane), queue length (veh), and TTS (veh.hour)

From the same initial mean speed in every segments of both network, we get the next mean speed in segment 4 link 1 network 1 is always higher than mean speed in the same segment in network 2. The comparation can be seen in Fig. 11. For traffic demand scenario 1, in the first network, average mean speed in segment 4 link 1 is 56.1882 km/hour, while average mean speed in the second network (network with off-ramp) is 54.3215 km/h. For traffic demand scenario 2 and 3, in the first network, average mean speed in segment 4 link 1 is 56.1824 km/hour, while average mean

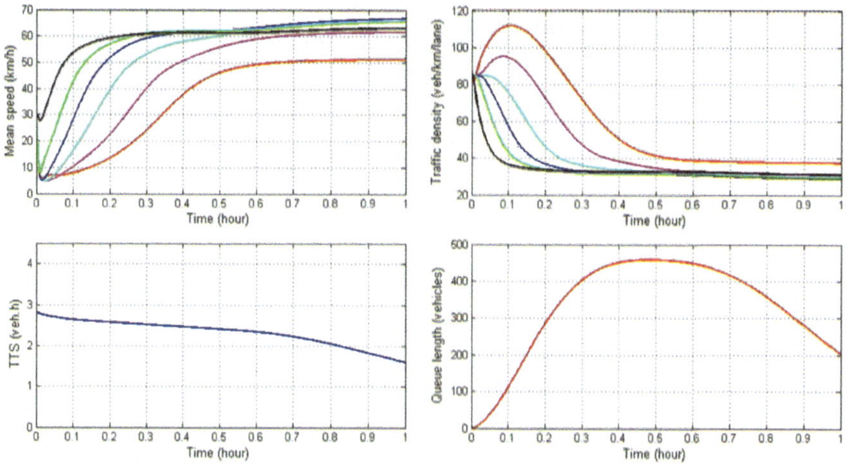

Fig. 7 Simulation result for the first traffic network (without off-ramp) with demand scenario 2: mean speed (km/hour), traffic density (veh/km/lane), queue length (veh), and TTS (veh.hour)

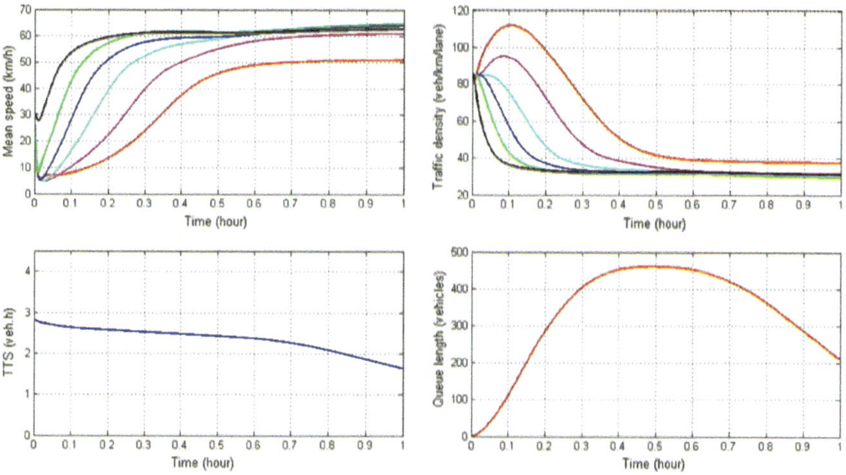

Fig. 8 Simulation result for the second traffic network (with off-ramp) with demand scenario 2: mean speed (km/hour), traffic density (veh/km/lane), queue length (veh), and TTS (veh.hour)

speed in the second network (network with off-ramp) is 54.3191 km/h, for parameter $\delta_{\text{off}} = 0.4$. Moreover, the speed reduction due to off-ramp also influences traffic density and total time spent (TTS). For the traffic demand scenario 1, TTS in network 1 is 747.1901 vehicles.hour, while TTS in network 2 is 753.8113 vehicles.hour. For the traffic demand scenario 2, TTS in network 1 is 842.4618 vehicles.hour, while TTS in network 2 is 849.083 vehicles.hour. For the last traffic demand scenario,

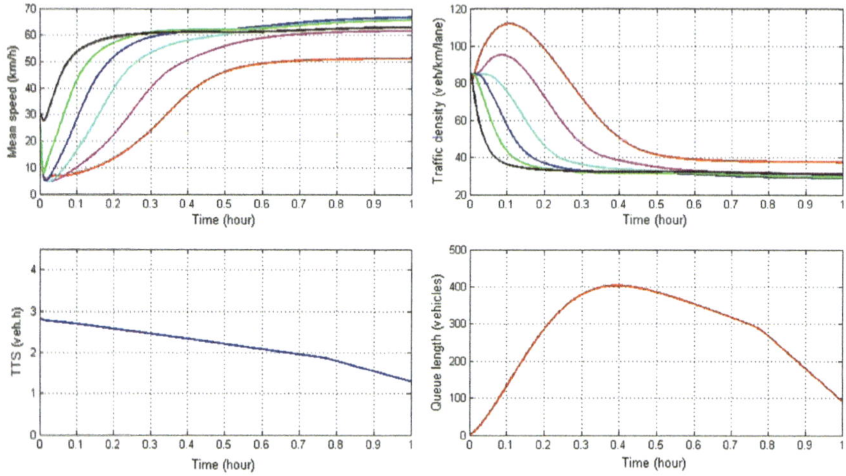

Fig. 9 Simulation result for the first traffic network (without off-ramp) with demand scenario 3: mean speed (km/hour), traffic density (veh/km/lane), queue length (veh), and TTS (veh.hour)

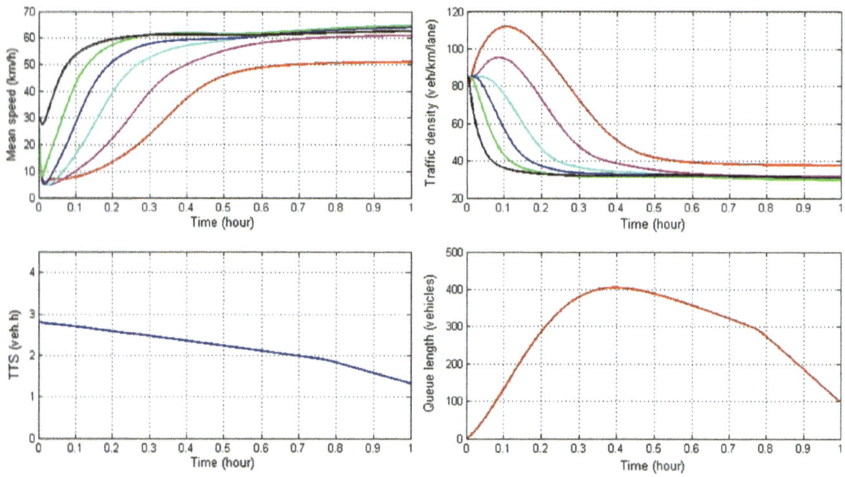

Fig. 10 Simulation result for the second traffic network (with off-ramp) with demand scenario 3: mean speed (km/hour), traffic density (veh/km/lane), queue length (veh), and TTS (veh.hour)

TTS in network 1 is 785.8521 vehicles.hour, while TTS in network 2 is 792.4733 vehicles.hour.

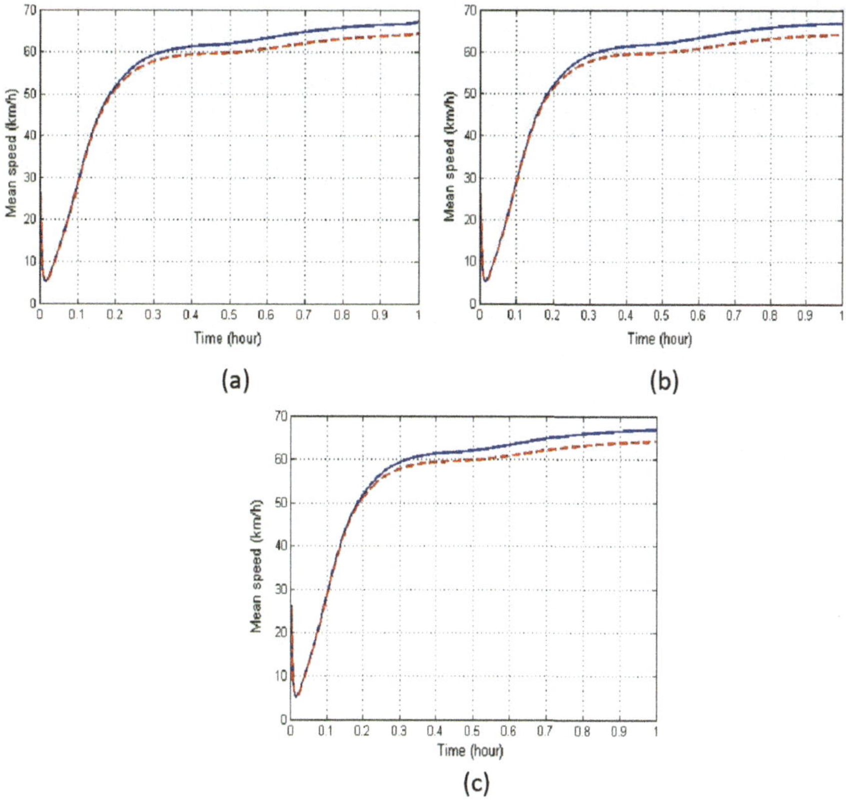

Fig. 11 The comparation between $v_{1,4}(k)$ in network 1 (blue curve) and in network 2 (red curve): **a** for demand scenario 1, **b** for demand scenario 2, **c** for demand scenario 3

5 Conclusion

The influence of off-ramp volumes to mean speed is discussed in this paper. The effect of existence of off-ramp volumes is reduction of speed in related segment. Numerical implementation as an illustration of speed reduction due to the off-ramp with some traffic demand scenarios has been carried out. This additional factor can be used to apply further implementation of METANET, for example, implementation which uses MPC for traffic control.

Acknowledgements The authors would like to gratefully acknowledge the funding from the International Journal Publication and Research Collaboration Grant, Directorate of Higher Education, Ministry of Education, Indonesia 2016 No. 1023/UN1-P.III/LT/DIT-LIT/2016.

References

1. Kotsialos, A., Papageorgiou, M., Diakaki, C.H., Pavlis, Y., Middelham, F.: Traffic flow modeling of large-scale motorway networks using the macroscopic modeling tool METANET. IEEE Trans. Intell. Transp. Syst. **3** (2002).
2. Lighthill, M.J., Whitham, G.B.: On kinematic waves, II. A theory of traffic flow on long crowded roads. In: Proceedings of the Royal Society 229A (1955)
3. Richards, P.I.: Shock waves on the highway. Oper. Res. **4**, 42–57 (1956)
4. Payne, H.J.: Models of freeway traffic and control. In: Simulation Council Proceedings (1971), pp. 51–61
5. Papageorgiou, M., Blosseville, J.M., Hadj-Salem, H.: Macroscopic modelling of traffic flow on the Boulevard Peripherique in Paris. Trans. Res. Part B **23B**(1), 29–47 (1989)
6. Messmer, A., Papageorgiou, M.: METANET: a macroscopic simulation program for motorway networks. Traff. Eng. Control **31**, 466–470 (1990)
7. Liu, S., Schutter, B.D., Hellendoorn, H.: Model Predictive Traffic Control Based on a New Multi-Class METANET Model. In: 19th World Congress of The International Federation of Automatic Control (2014)
8. Technical University of Crete, Messmer, A.: METANET—a simulation program for motorway networks. Technical University of Crete (2012)
9. Bellemans, T., Schutter, B.D., Moor, B.D.: Models for Traffic Control. J. A **43**, 13–22 (2002)
10. Hegyi, A.: Model predictive control for integrating traffic control measures. Ph.D. Thesis, Delft University of Technology (2004).
11. Spiliopoulou, A., Kontorinaki, M., Papamichail, I., Papageorgiou, M.: Real-time route diversion control at congested motorway off-ramp areas—part I: user-optimum route guidance. In: 16th International IEEE Annual Conference on Intelligent Transportation Systems (2013a)
12. Spiliopoulou, A., Kontorinaki, M., Papamichail, I., Papageorgiou, M.: Real-time route diversion control at congested off-ramp areas—Part II: Route guidance versus off-ramp closure. In: EWGT2013—16th Meeting of the EURO Working Group on Transportation (2013b)
13. Spiliopoulou, A., Papageorgiou, M., Herrera, J.C., Muñoz, J.C.: Real-time merging traffic control at congested freeway off-ramp areas. In: 95th Annual Meeting of the Transportation Research Board (2016)
14. Groot, N., Schutter, B.D., Hellendoorn, H.: Integrated model predictive traffic and emission control using a piecewise-affine approach. IEEE Trans. Intell. Transport. Syst. **31** (2013)
15. Hegyi, A., Schutter, B.D., Hellendoorn, H.: Model predictive control for optimal coordination of ramp metering and variable speed limits. Transport. Res. Part C **13**, 185–209 (2005)
16. Groot, N., Schutter, B.D., Zegeye, S., Hellendoorn, S.: Model-based predictive traffic control: piecewise-affine approach based on METANET. In: 18th IFAC World Congress, Milan, Italy (2011), pp. 10709–10714

An Adaptive Queue-Length Estimator Based on Stochastic Hybrid Model

Herman Y. Sutarto and Endra Joelianto

Abstract This paper presents a method for particle filter (PF)-based joint state and parameter estimation for a stochastic hybrid model of queue-length dynamics for urban traffic. The model represents continuous variables, including the stochastic traffic flow rates entering and exiting the queue during successive green and red periods, and discrete event variables including the green/red switching and the traffic flow modes (free flow versus congested). A particle filtering approach is used for jointly estimating the transition matrix of the traffic modes, the parameters of the autoregressive (AR) model of the traffic flow rates, as well as for estimating and predicting the traffic flow rates and the queue length themselves. Obtaining computationally efficient PFs for this requires several improvements to existing techniques. We use an optimal tuning kernel smoothing approach to estimate the parameters of the first-order AR model in combination with the Dirichlet distribution approach to update the parameters of the transition probability matrix of a first-order Markov chain. The proposed technique is validated and evaluated by comparing the queue size estimations with synthetic data generated by a VISSIM traffic micro-simulator. The results of this analysis confirm that the hybrid model together with the PF parameter estimator gives satisfactory results, properly capturing the evolution in time of queue length and traffic flow. This algorithm may therefore be included in the feedback loop of adaptive traffic signal controllers.

Keywords Stochastic hybrid system · State-parameter estimation · Particle filter · Data fusion · Urban traffic networks

H. Y. Sutarto
Institut Teknologi Harapan Bangsa, Bandung 40132, Indonesia
e-mail: hy.sutarto@rce.co.id

E. Joelianto (✉)
Faculty of Industrial Technology, Institut Teknologi Bandung, Bandung 40132, Indonesia
e-mail: ejoel@tf.itb.ac.id

E. Joelianto et al. (eds.), *Cyber Physical, Computer and Automation System*, Advances in Intelligent Systems and Computing 1291,
https://doi.org/10.1007/978-981-33-4062-6_12

1 Introduction

Implementing advanced traffic signal control strategies requires real-time data processing for online estimation of the key state variables as well as of the model parameters, neither of which can in practice be measured directly with sufficient accuracy. An extensive literature is available on online traffic flow state estimation for freeway traffic and for arterial roads, often using sub-optimal Bayesian filters and particle filters [1]; online joint state-parameter estimation for urban networks on the other hand has not received much attention yet.

Traffic in urban networks is characterized by stop-and-go phenomena resulting from green/red switching and from complicated interactions between conflicting traffic streams. In [2] and [3], we showed that the variability of urban traffic flow at traffic lights can be approximated by an autoregressive process (AR), whose model parameters evolve over time depending on the mode of traffic operation (free-flowing or congested/faulty); the unobservable discrete event model of the mode of operation is a first-order Markov chain. Hence, we model the queue-length dynamics as a stochastic hybrid model, taking into account the strong interaction between traffic flow and traffic light.

Coordination between traffic lights at neighboring intersections plays a major role in reducing congestion and avoiding gridlock in an urban network. By coordinating actions of local intersection controllers efficient and effective dynamic traffic management can be achieved. Coordination requires anticipation of queue sizes at neighboring intersections, and this anticipation is only possible if a good model is available for predicting future traffic behavior, and in particular future queue sizes. In order to use a dynamic queue-length model, one must estimate the parameters of this model through online state-parameter estimation. Automatic adaptation to changing system condition thus becomes possible, provided we can estimate the model parameters online with sufficient accuracy.

The main contributions of this paper are using a stochastic hybrid model describing queue-length dynamics for state-parameter estimation of that model along with a Dirichlet distribution to generate samples that allow reliable estimates of transition probabilities. By comparing the estimated and predicted queue sizes to queue sizes obtained using synthetic queue-length data as generated by a VISSIM traffic simulator, we show that the joint state and parameter estimation tools proposed in this paper exhibit good performance for realistic scenarios. The reason we use synthetic data for the validation of our estimation algorithm is that it is very hard to obtain reliable real data on both the flow rates and the queue sizes, due to the limitations of the traffic sensors of the partners from which we tried to obtain real traffic data.

The present paper proposes an extension of adaptive particle filters for combined state-parameter estimation of the hybrid model of queue-length dynamic. This adaptive SIR filter achieves adaptation by using an online optimization algorithm based on Kullback–Leibler (KL) divergence. Originally, this adaptive SIR filter was proposed by Tulsyan [4] for a class of nonlinear models. Since the hybrid model depends on the different traffic modes, the transition probabilities matrix (TPM) is also a parameter to be estimated. We use the Dirichlet distribution to estimate this TPM, searching for reliable transition probabilities rather than applying a multi-normal kernel smoothing as proposed by Carvalho [5]. The combination of adaptive PF with Dirichlet-based kernel smoothing leads to an efficient estimation of the TPM.

The remainder of this paper is organized as follows. Section 2 explains the problem formulation including the piecewise affine model of the queue length, while Sect. 3 explains the jump Markov model structure. Section 4 discusses online Bayesian estimation; the performance evaluation is discussed in Sect. 5. Section 6 summarizes the findings of the paper.

2 Stochastic Hybrid Model of Queue Length

This section introduces a stochastic hybrid model for the queue-length dynamic along one particular approach route to a signalized intersection and explains how the parameter estimation problem treated in this paper fits into the overall traffic control problem. This shows how this model is useful in controlling the operation of a signalized intersection.

By using online measurement data of the existing traffic flow for online parameter estimation, we build a dynamic traffic flow model and use this model for adaptively predicting future values of the arrival flow rate $\boldsymbol{\lambda}(\mathbf{t})$ and the departure flow rate $\boldsymbol{\mu}(\mathbf{t})$. A good model should have the capability to cope with variation of these traffic flow rates and should be usable for developing queue-length-based traffic control. Indeed, the evolution of the queue length is defined by

$$Q_{\mathrm{L}}(t) = Q_{\mathrm{L}}(t_0) + \int_{t_0}^{t} (\lambda_{\mathrm{L}}(\tau) - \mu_{\mathrm{L}}(\tau))\mathrm{d}\tau \tag{1}$$

Equation (1) describes the evolution over time of queue length (in the further analysis and simulation experiments we use data for L4 see Fig. 1). The traffic flow rate $\lambda_{\mathrm{L}}(t)$ (resp. $\mu_{\mathrm{L}}(t)$) of the arrival stream of vehicles (resp. the departure stream of vehicles) for L4 is measured in veh/s. In general, traffic flow $\alpha(t)$ (the generic notation for $\lambda_{\mathrm{L}}(t)$ and $\mu_{\mathrm{L}}(t)$ used further on) describes the number $\alpha(t)\mathrm{d}t$ of vehicles passing a given location during the time interval $[t, t + \mathrm{d}t)$ (in order to avoid working with large integers we fluidize our traffic variables, approximating integer numbers of vehicles by real number, see [2]. The traffic flow $\alpha_k = \alpha(t_k) = N_k/(t_{k+1} - t_k)$ (where N_k counts the number of vehicles passing the given location in the interval $[t_k, t_{k+1})$) is measured at an increasing sequence of discrete instants in time t_k. The fluid flow approximation implicitly assumes that the traffic flow $\alpha(t) = \alpha_k$ for $t \in [t_k, t_{k+1})$ remains constant during the interval $[t_k, t_{k+1})$.

We assume that as long as the mode of operation of the traffic—either free-flowing or congested—remains unchanged α_k evolves as a first-order autoregressive (AR) random process. The parameters of this AR process however take different values depending on the mode s_k of traffic operation. We assume that the interval between

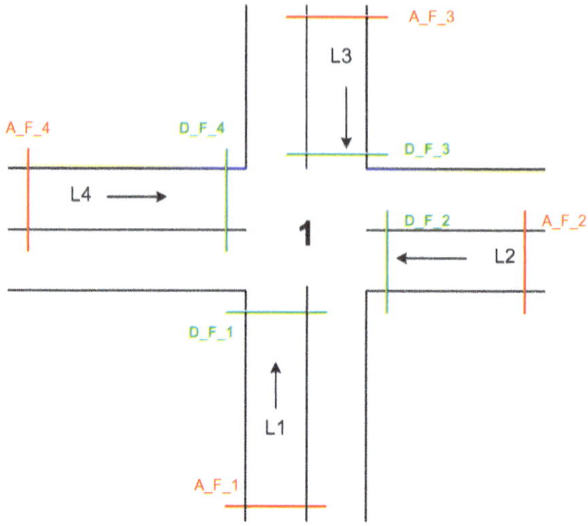

Fig. 1 Single intersection with incoming lanes

successive events is geometrically distributed, so that the mode $s_k(\alpha_k)$ is represented by a first-order Markov chain. The hybrid stochastic process $(\alpha_k, s_k(\alpha_k))$ is called a jump Markov model. The transition probability matrix Π of the mode switching system must be estimated, as well as the parameters of the AR model for each mode.

The design of a good traffic controller requires a good model describing the evolution of the arrival and departure traffic flows, λ_k and μ_k. In this paper, unlike the model in [2], we split up the arrival traffic flow into two parts: λ_{2k} during $T_g = t_{2k+1} - t_{2k}$ (green period) and λ_{2k+1} during $T_r = t_{2k+2} - t_{2k+1}$ (red period), please see Fig. 2. We also define departure traffic flow μ_{2k} ($\mu_{2k+1} = 0$ as there can be no departures during the red). Similar to what has been done in [2] one can validate,

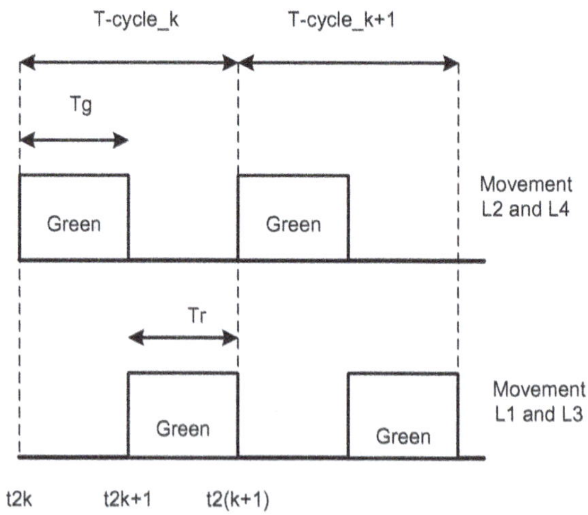

Fig. 2 Signal traffic sequence

using real data, that the traffic flow rates λ_{2k}, λ_{2k+1}, and μ_{2k} are properly modeled by an AR process, with the parameters that depend on the mode s_k of traffic operation (free-flowing and congested flow are considered only here, a simplification of the three-mode model of [2]).

This model defines the queue sizes at the times $t_{k,}$ as functions of λ_{2k}, λ_{2k+1}, and μ_{2k}:

at t_{2k+1}:

$$q_{2k+1} = \max\left[q_{2k} + (\lambda_{2k} - \mu_{2k})T_g, 0\right] \tag{2}$$

at t_{2k+2}:

$$q_{2k+2} = q_{2k+1} + \lambda_{2k+1}T_r \tag{3}$$

In compact form, this can be written as:

$$q_{2k+2} = \max\left[q_{2k} + (\lambda_{2k} - \mu_{2k})T_g, 0\right] + \lambda_{2k+1}T_r \tag{4}$$

The hybrid model $(q_k, (\lambda_{2k}, s_k((\lambda_{2k})), (\lambda_{2k+1}, s_k((\lambda_{2k+1})), (\mu_{2k}, s_k((\mu_{2k})))$, allows us to predict the evolution of traffic flows and of queue sizes a few cycles ahead. This anticipation is essential for feedback control design of traffic light operation.

3 Jump Markov Model Structure

In this section, we describe in detail the stochastic hybrid model $(\alpha_k, s_k(\alpha_k))$, also called further the jump Markov model, for a generic traffic flow rate α_k (to simplify the notation, we denote $s_k(\alpha_k) = s_k$ further on in this section). During a period of time when remains unchanged, this variable is modeled by a first-order autoregressive (AR) model:

$$\alpha_{k+1} = (\beta(s_k) + \gamma(s_k)\alpha_k + w_k) \tag{5}$$

where w_k, $k = 1, 2 \ldots$ is an independently identically distributed sequence of zero mean random variables with variance $\sigma^2(s)$, $\beta(s)$, and $\sigma^2(s)$ mode dependent parameter to be identified.

From time to time the mode of operation s_k will change, due to external changes of the inflow rate, or due to incidents that make the operation more or less efficient, or due to randomness. We assume that the mode changes occur at time t_k when the traffic flow rate α_k is updated. We consider two different values:

- $s_k = 1$ is the desirable mode of operation with traffic flowing with little interference between successive vehicles;

- $s_k = 2$ is the congested mode when vehicles hinder each other significantly, and the system operates inefficiently;

We assume that the mode process s_k can be modeled by a Markov chain, i.e., at each time t_k the mode variable $s_{k-1} = i$ changes randomly to the value $s_k = \text{j}$ with a probability $\pi_{ij} = \text{Prob}\ (s_k = j | \ s_{k-1} = i,\ s_{k-2,}\ s_{k-3},\ldots)$ dependent only on the most recent value s_{k-1}, not on states further in the past. This Markovian assumption is reasonable for systems that exhibit abrupt random changes that can be caused by many different causes such as the onset of the rush hour, accidents, and random perturbations. More complicated models would not be useful anyway since it would be too difficult to estimate good parameter values.

Equation (5) with its interpretation $\alpha(t) = \alpha_k$ for $t \in [t_k, t_{k+1})$, together with the Markov chain model for the mode s_k provides us with a complete mathematical model of traffic flow. The parameters of this model have to be estimated online. In total—for a single traffic flow α_k—there are eight parameters to be estimated:

$$\theta_\alpha(s_k) = \left\{\beta_\alpha(s_k),\ \gamma_\alpha(s_k),\ \sigma_\alpha^2(s_k),\ \pi_\alpha\right\} \tag{6}$$

where α is a generic flow (it can be λ_{2k}, λ_{2k+1}, μ_{2k},):

- For each mode $s_k \in \{1,2\}$, the AR model in Eq. (3) has three parameters, $\beta(s)$, $\gamma(s)$, and $\sigma^2(s)$.
- The transition matrix $\Pi_\alpha = (\pi_{ij})_\alpha$ of the Markov chain has 2 rows of 2 elements, each summing to 1, leaving one free parameter per row.

Equation (6) implies that one treats s_k (λ_{2k}), s_k (λ_{2k+1}), and s_k (μ_{2k}) as three independent processes. By assuming this, for example, it is allowed that the arrival flow rates in the red and the green period of the cycle correspond to two different traffic condition, e.g., due to condition at adjacent intersections. Surely, the processes are in practice strongly correlated, but for the time being, we simplify by assuming that each process is independent.

The data used in the recursive Bayesian described in Sects. 4 and 5 (also the data that an online traffic controller can use) are obtained by counting the observed number of vehicles $y_k = N_k/(t_{k+1} - t_k)$ passing the location to which the model is applied during the interval $t \in [t_k, t_{k+1})$:

$$y_k = \alpha_k + \eta_k \tag{7}$$

where n_k, $k = 1, 2, \ldots$ is a sequence of independent zero mean random variables with Gaussian distribution with variance ${\sigma_n}^2$ (representing observation errors due to false detections and missed detections during the often inaccurate counting provided by video cameras or magnetic loops). This variance should be chosen realistically. For the results in this paper, we used a random walk approach to estimate the sample variance using historical data for $(y_{k+1} - y_k)$.

We split up the vehicle counts in a red and a green period as a measurement period because: (a) the queue-length model in (4) has two periods, one covering the start of the cycle to the end of green, the other between the end of green and the end of the cycle, (b) if we look at the values of mean and variance of the traffic flow generated by the VISSIM simulator (simulating a signalized intersection during one hour with a constant arrival rate; for details see Sect. 5) with a cycle length 80 s and green period 35 s. We select update intervals of around 35 s (typical green period) that provide low variance.

4 Online Bayesian State-Parameter Estimation for Stochastic Hybrid System

In this section, we develop a complete approach to online Bayesian joint state and parameter estimation for hybrid stochastic systems $\{x_k\} = \{q_k, \lambda_{2k}, \lambda_{2k+1}, \mu_{2k}, s_k(\lambda_{2k}), s_k(\lambda_{2k+1}), s_k(\mu_{2k})\}$ as defined by (4, 5), using an extended state vector representation with artificial dynamics for its parameters θ_k as defined by (6). We discuss the Bayesian framework for particle filter (PF) and its extension to a hybrid system using the adaptive PF technique. Combined state-parameter estimation will be discussed through joint posterior distribution.

The *state model*, for the augmented process $\{\chi_k\} = \{x_k, \theta_{k.}\}$and this is combined with *the observation model* (7) specifying the probability distribution of the data y_k given the underlying state x_k. This description of the evolution of the augmented state, together with initial conditions, provides the necessary information to predict the queue sizes a few cycles ahead.

Originally, adaptive PF was proposed solely for state estimation of hybrid systems, relying on prior knowledge of the parameters $\theta_k(\alpha_k)$, including the transition probabilities Π_α. Since we do not know the parameters of our traffic model, we also must identify these parameters of the hybrid system models, using the online data observed during the plant operation, thus achieving automatic adaptation of the state predictors to the changing system condition. We achieved this by finding a good compromise between computational efficiency and accuracy for estimating the traffic state and for each mode s:

a. The parameters of the AR model;
b. The transition probability matrix Π_α.

Problem (a) will be discussed in Sect. 4.2 using parameter tuning for kernel smoothing, while problem (b) will be solved in Sect. 4.3 using a Dirichlet distribution.

4.1 State Estimation of Hybrid System

In [6], it is shown that the adaptive PF technique (see [2, 3, 7]) is more efficient in computational complexity than interacting multiple model particle filtering (IMMPF) proposed by [8] for state estimation in hybrid system. State estimation based on hybrid system calculates the PDF: $p(x_k, s_k \mid Y_k)$, where s_k is the mode of the system at time k treating the mode as an unknown system parameter. Since at each time step, the system only follows one mode, it is reasonable to assume that it is actually only following the most likely mode $\hat{s}_k$:

$$\begin{aligned} p(x_k, s_k|Y_k) &= p(s_k|Y_k)p(x_k|s_k, Y_k) \\ &\sim p(\hat{s}_k|Y_k)p(x_k|\hat{s}_k, Y_k) \end{aligned} \tag{8}$$

$$= p(\hat{s}_k|Y_k)p_{\hat{s}_k}(x_k|Y_k) \tag{9}$$

Since $p(\hat{s}_k| Y_k)$ is a constant in (9), its effect will be absorbed in the normalization constant and will not affect the basic PF algorithm. In [2], it has been shown that this approach, substituting s_k by its most likely value, leads to sufficiently accurate results.

4.2 Parameter Tuning for Kernel Smoothing

If both state and model parameters must be estimated simultaneously, Bayes' rule must calculate the following joint posterior distribution:

$$\begin{aligned} &p(x_k, \theta_\alpha(s_k)|Y_k) \sim p(y_k|x_k, \theta_\alpha(s_k)) \\ &p(x_k|\theta_\alpha(s_k), Y_{k-1})p(\theta_\alpha(s_k)|Y_{k-1}) \end{aligned} \tag{10}$$

where $\theta_\alpha(s_k)$ is the vector of model parameters in Eq. (6). The joint state and parameter estimation of the AR process is performed by adding artificial dynamics to the parameters (e.g., a random walk) in the extended state vector, containing both states and parameters. Liu in [8] shows that the Monte Carlo approximation $\{\theta_k^j, w_k^j\}$ to $p(\theta| Y_k)$ has mean $\overline{\theta}_k$ and variance V_k. Hence, the evolution $\theta_{k+1} = \theta_k + \varsigma_{k+1}$, the implied prior $p(\theta_{k+1}| Y_k)$ has correct mean $\overline{\theta}_k$ but variance $V_k + W_{k+1}$, where W_{k+1} is variance of the innovation ς_{k+1}. The random walk approach leads to an increase in the covariance. A natural approach to reducing the covariance is to use kernel smoothing with a properly tuned smoothing factor. We can approximate the last term in Eq. (10) by mixture of particles and weights $\{\theta_k^j, w_k^j\}$:

$$p(\theta_\alpha(s_k) \mid Y_{k-1}) \approx \sum_{i=1}^{N} w_{k-1}^i \, G\left(\theta_k \mid m_{k-1}^i, h^2 V_{k-1}\right)$$

The kernel location $m_{k-1}{}^i$ is specified by a shrinkage rule that forces the particles to be closer to their mean. A detailed discussion of why this avoids the increase in the covariance matrix applied to the case of traffic flow estimation can be found in [9].

The optimal selection of the kernel parameter $h_k \in [0,1]$ maximally reducing the over-dispersion in the PF remains a difficult problem. The current practice for tuning of h_k is ad hoc. Liu [9] suggested choosing $h_k = 0.1$, whereas Chen et al. [10] select h_k as the optimum for a historical dataset, and then applies this choice for future batches. These ad hoc rules define a constant h_k, for which optimality cannot be established w.r.t the online data.

We develop an optimal tuning rule for h_k based on an online optimization [3] for stochastic hybrid systems. This tuning rule minimizes the Kullback–Leibler (KL) divergence $D(h_k)$ between $p(\chi_k \mid Y_{k-1})$ and the target posterior density $p(\chi_k \mid Y_k)$ at each sampling time. The objective of this optimizer is not only to tune h_k, but also to project the particles sampled from $p(\chi_k \mid Y_{k-1})$ into the region of high posterior density $p(\chi_k \mid Y_k)$. This enables adaptation of the SIR filter for combined state-parameter estimation. Proposition 1, adopted from [4], provides an optimal tuning rule for controlling the kernel width and for making an SIR filter adaptive.

Proposition 1 The optimal value $h^*{}_k$ of the parameter h_k at time t_k that minimizes the KL divergence $D(h_k)$ between the $p(\chi_k \mid Y_{k-1}))$ and target posterior density $p(\chi_k \mid Y_k)$ is:

$$\begin{aligned} h_k^* &= \underset{h_k \in [0,1]}{\arg\min} \left[-\sum_{i=1}^{N} w_k^i \log_{10}\left[\omega_k^i\right] \right] \\ &= \underset{h_k \in [0,1]}{\arg\min}\, \hat{D}(h_k) \end{aligned} \tag{11}$$

where: $\hat{D}\,(h_k)$ is a sequential Monte Carlo (SMC) estimate of $D\,(h_k)$. Note that the dependence of $\hat{D}\,(h_k)$ on h_k can be established from the transition of state in Eq. (5) and parameter and also shrinkage equation. $\{w^i{}_k\}^N{}_{i=1}$ and $\{\omega^i{}_k\}^N{}_{i=1}$ are the particle weights given in (8) and (12), respectively:

$$\omega_k^i = \frac{w_k^i p\left(y_k \mid Z_{k|k-1}^i\right)}{\sum_{i=1}^{N} w_k^i p\left(y_k \mid Z_{k|k-1}^i\right)} \tag{12}$$

where:

$$\left\{Z_{k|k-1}^i\right\}_{i=1}^N \sim \tilde{p}(z_k \mid y_{1:k-1})$$

$$z_k \triangleq \{x_k, \theta_k\}$$

4.3 Dirichlet Distribution

For estimating the transition probabilities of the mode process s_k, the parameters are updated by generating samples of the modes according to the Dirichlet distribution. The ith row $\{p_{i1}, \ldots, p_{ij}\}$ of an estimate of the TPM Π_i always has row sum equal to 1. Let Π_i be random variables: $\Pi_i \sim D(\lambda_{i1},\ldots, \lambda_{ij})$, where D denotes a Dirichlet distribution. It is known that (see [10]), the updated distribution of $P|S_t$ is also a Dirichlet distribution, where $S_k = \{s_1, s_2, \ldots, s_k\}$:

$$P|S_k \sim D\left(\lambda_{i1+n_{i1}}, \lambda_{i2+n_{i2}}\right) \tag{13}$$

where n_{ij} is the number of one-step transitions from i to j in sample S_k. Therefore, we propose the Dirichlet distribution to search reliable transition probabilities rather than applying a multi-normal kernel smoothing as proposed by [5]. In contrast to the method in [11], our approach does not enforce that all modes be visited since our approach uses the method considering only the most likely mode.

To update the Dirichlet distribution, we use z_p gamma random variables for generating the samples.

The vector Π_i $(1 \leq i \leq j)$ can now be simulated from (13) by letting:

$$p_{i1} = \frac{z_1}{\sum_{p=1}^{j} z_p} \ldots p_{i2} = \frac{z_2}{\sum_{p=1}^{j} z_p} \tag{14.a}$$

$$z_p \sim \text{Gamma}\left(\lambda_{ip} + n_{ip}, 1\right) \tag{14.b}$$

5 Online Bayesian Performance Evaluation

To test the performance of the proposed estimation and prediction algorithm for the queue length in a field experiment is difficult. We have not been able to find data where simultaneously traffic flow and queue length where recorded with sufficient accuracy over the cycles of the traffic light, as we consider in our model. Fortunately, the advanced computational power and the flexibility that state-of-the-art computer-based simulation software offers make it possible to validate our algorithm using the VISSIM traffic micro-simulator. We use VISSIM to generate synthetic traffic data implementing a detailed version of the hybrid dynamic model introduced in Sects. 2 and 3, and generating the noise as required for realistic representation of traffic irregularity, mode changes, and measurement noise. The output of such a simulation run provides the noisy data about traffic flow rates and queue lengths under various conditions, for the time intervals corresponding to the cycle of the traffic lights, used as input for our joint state and parameter PF estimator as described in Sect. 4. In order to achieve fast results applicable to online implementation, our PF particles are generated by a simplified version of our hybrid model. Hence, we can compare the queue length obtained via microsimulation with the estimations and predictions obtained via the PF estimator.

Our simulation experiment represents the typical situation at one intersection, with two-way approach roads of length 300 m each, the distances between the source generating the traffic flow and the first measurement station, resp., the second station being 0 m, resp., 300 m; the second station locates is located at the stop line. The speed distribution in this validation experiment is 35–65 kph, chosen on the basis of prior experience with VISSIM. The average distance between stopped cars and also between cars and stop lines, signal heads, and so forth is 2.0 m, uniformly distributed in the interval [1.0 m 3.0 m]. The lateral behavior parameters allow overtaking wherever legal and when traffic flow conditions allow this. We assume that one vehicle occupies the full width of one lane in VISSIM. The type of vehicles is mixed traffic consisting of cars and motorcycles. The minimum (maximum) length of a car is 4.1 m (4.7 m). The length of a motorcycle is 1.4 m. For the sake of simplicity, the traffic model implemented in the PF algorithm does not distinguish between cars and motorcycles, and the length of any vehicles is approximated by the average $(4.4\text{ m} + 1.4\text{ m})/2 = 2.9$ m.

The traffic signals in our simulation operate according to a fixed cycle. The cycle length was set to 80 s with green duration is 35 s, amber duration is 3 s and all-to-red is 2 s. The vehicle flow of generation origin is 0.16 veh/sec and the simulation time interval is 3600 s. In VISSIM, It is noted that the number of vehicles is counted by meter according to the classification of the vehicle whether it be a car or a motorcycle. In Table 1, the average traffic flow is ≈ 0.2 veh/sec and this value is computed based on the average length of vehicles 2.9 m, meaning that one does not distinguish between cars and motorcycles. This approach would be used in PF algorithm.

In performance evaluation purposes, it is considered the VISSIM microscopic approach for constructing the actual queue length. By using the microscopic simulation approach which the current queue length is measured upstream from the intersection at every time step, including at the switching times when we need the queue size in order to compare to what our model predicts. The actual VISSIM queue length is recorded by putting the queue counter in the front of stop line *at the end of every red light* (cycle-by-cycle). This data is directly generated in VISSIM as microsimulation,

Table 1 RMS performance

Type\N	500		5000		20,000	
	Pred-1	Pred-2	Pred-1	Pred-2	Pred-1	Pred-2
VISSIM	296.5	310.5	252.3	253.3	239.6	244.5

meaning that the VISSIM do distinguish between length of car and of motorcycle to define the queue length (meter). VISSIM outputs is the length of the queue at time t_k measured in meters, and this output is a result of multiplying the number of vehicles by length of each vehicle.

The queue-length model in Eq. (4) and PF-based joint state-parameter estimation will be validated by using VISSIM traffic simulator. In this PF approach, the queue length is defined based on the cumulative numbers of vehicles traversing sensor locations (in this case study AF-4 and DF-4) are recorded during successive phases of traffic signal, providing data for calculating time series data $\lambda_{2k}.T_g$, $\lambda_{2k+1}.T_r$, and $\mu_{2k}.T_g$ as shown in Eq. (4). In VISSIM, it is considered that each vehicle can be classified into car or motorcycle corresponding to the length of each vehicle. In order to generate particles fast enough for online implementation of the PF, the PF does not distinguish between cars and motorcycles, and the length of any vehicles is approximated by its average value 2.9 m.

Figure 3 shows that the PF queue-length estimator and predictor gives results close to the actual VISSIM queue length. The biggest differences are in index 23; detailed analysis of the measurement data indicates that around that time the most vehicles that pass detector are cars, causing the PF simplification of ignoring the difference between cars and motorcycles to be significant. The implementation of the joint state and parameter estimation with optimal tuning kernel smoothing under a certain initial condition and $N = 500$ samples give a good result in terms of queue length and traffic flow prediction (1-cycle/ 2-cycle ahead) as shown in Figs. 3 and 4.

Increasing the number N of samples of course has even better performance of predicting the queue length 1-cycle ahead and 2-cycle ahead but not so much better as to make it worthwhile, given the increase in computational cost. Prediction of traffic flow over one or two-cycle ahead is performed by using:

$\mathrm{FP}_1(k + 1) = (\mathrm{FE}\ (k))$*(the right-term of Eq. (5)).

$\mathrm{FP}_2(k + 2) = (\mathrm{FP}_1(k + 1))$*(the right-term of Eq. (5)).

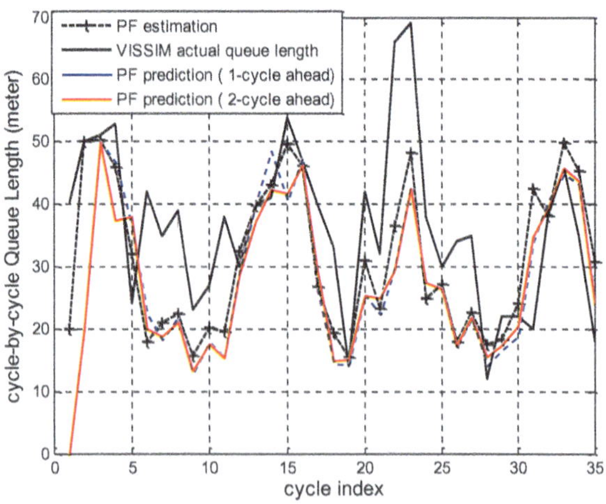

Fig. 3 Queue-length prediction: $N = 500$

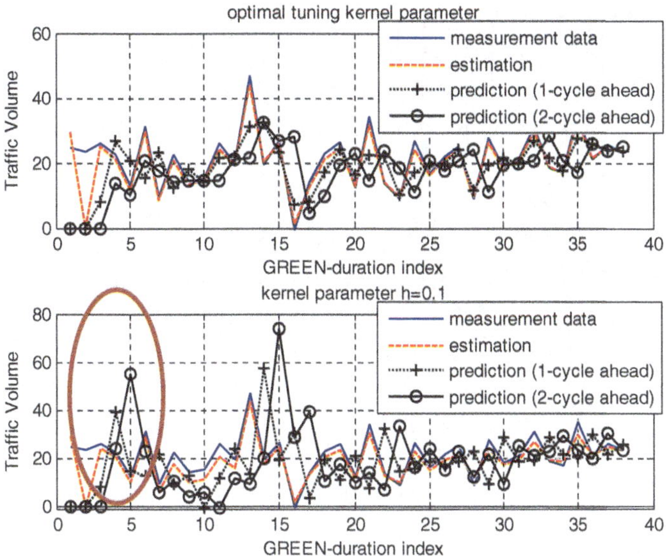

Fig. 4 Arrival flow estimation and prediction (upper) with optimal tuning h (lower) with $h = 0.1$

where: FE(k) is a flow estimation at kth cycle; $\mathrm{FP}_1(k+1)$ is flow prediction at $(k+1)$th cycle based on FE(k), and FP_2 $(k+2)$ is flow prediction at the $(k+2)$th cycle based on $\mathrm{FP}_1(k+1)$.

The right-term of Eq. (5) is determined using the identified parameters $\theta(s_k) = [\theta_\alpha, \Pi]$ in Eq. (6). We show in Table 2 the root mean square (RMS) error as a measure of performance, comparing "real" (albeit in our experiment simulated) against the predicted values generated by PF.

Selecting optimal parameters, like choosing the appropriate number N of particles or the optimal tuning of the kernel parameter, are important issues in this joint state-state-parameter estimation, given that the algorithm is computationally much more demanding than for state estimation only. The optimal compromise between performance and computational load for online applications is a topic for further research. Figure 4 compares the PF performance as a predictor, when using kernel parameter $h = 0.1$ as suggested by Liu [9] (lower part) versus the better prediction results using the optimal smoothing parameter, using the approach explained in Sect. 4.2

In order to obtain a good prediction, it is necessary to have an accurate estimation, and an accurate model describing the future evolution of the predicted variable. Figure 4 shows that the online PF with optimal tuning kernel smoothing is able to estimate the parameters of the AR process as well as transition probabilities Π. Figure 5 shows the evolution of mode changes of the departure flow data by choosing *shape* parameter of gamma distribution $\lambda = 0.5$. While our algorithm detects most of the time the most likely mode, it is clear that further improvement is needed for accurate mode estimation.

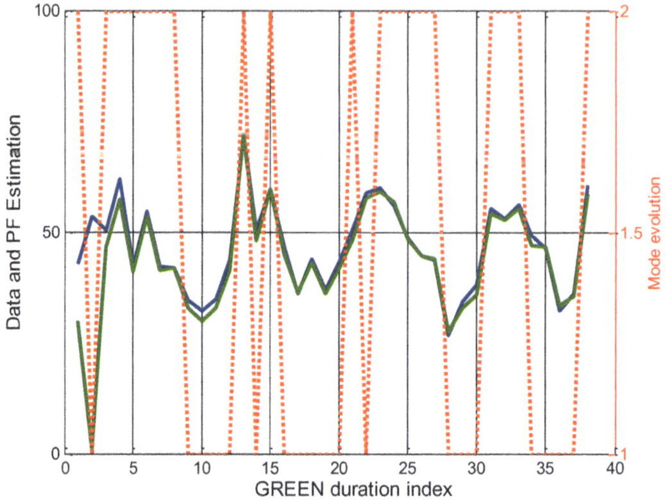

Fig. 5 Mode evolution of the departure flow

6 Conclusion

This paper presented a technique for joint parameter and state estimation for a stochastic hybrid model of the queue-length dynamic and its application to queue-length estimation and prediction at signalized intersections in urban traffic networks. The key of this method was to look at the mode as an unknown system parameter. The system was assumed to follow the dynamics of this most likely mode.

In estimating the parameters of the model, it was obtained significant improvement in performances by using the optimal tuning kernel smoothing technique while the estimation of the transition probability matrices was improved by using the Dirichlet. The proposed method was validated in terms of traffic flow and queue length. The case of synthetic data of queue length generated by VISSIM traffic simulator was used to show that online joint state-parameter estimation provides satisfactory queue-length estimation and prediction, and correctly captures the random variation of the traffic flow. The prediction results of traffic flows and queue length indicated that this proposed technique can be used to develop good anticipating traffic controllers which are an interesting topic to further research work.

Acknowledgements The authors would like to express gratitude to the Ministry of Research and Technology /BRIN under Fundamental Research Grant 2020, Indonesia and Institut Teknologi Bandung, Bandung, Indonesia.

References

1. Mihaylova, L., Boel, R, Hegyi.: A freeway traffic estimation within particle filtering framework. Automatica **43**(2), 290–300 (2007)
2. Sutarto, H.Y., Boel, R.K., Joelianto, E.: Parameter estimation for stochastic hybrid model applied to urban traffic flow estimation. IET Control Theo Appl **9**(11), 1683–1691 (2015)
3. Sutarto, H.Y., Joelianto, E., Nugroho, T.A.: Developing a stochastic model of queue length at a signalized intersection. Int. J. Adv. Sci. Eng. Inf. Technol. **7**(6), 2183–2188 (2017)
4. Tulsyan, A., Huang, B., Gopaluni, R.B., Forbes, J.: On simultaneous on-line state and parameter estimation in non-linear state-space models. J. Process Control **23**(4), (2013).
5. Carvalho, C.M., Lopes, H.F.: Simulation-based sequential analysis of Markov switching stochastic volatility models. Comput. Stat. Data Anal. **51**, 4526–4542 (2007)
6. Xue, Y., Runolfsson, T.: Efficient estimation of hybrid systems with application to tracking. Int. J. Syst. Sci. **43**(12) (2012)
7. Tafazoli, S., Sun, X.: Hybrid system state tracking and fault detection using particle filters. IEEE Trans. Control Syst. Technol. **14**(6), (2006)
8. Blom, H.A.P., Bloem, E.A.: Exact bayesian and particle filtering of stochastic hybrid system. IEEE Transact. Aerosp. Electron. Syst. **43**(1), 55–70 (2007)
9. Liu, J., West, M.: Combined parameter and state es-timation in simulation–based filtering, In: Arnaud, D., et al. (ed.) Sequential Monte Carlo Methods in Practice. Springer, New York (2001)
10. Chen, T., Morris, J., Martin, E.: Particle filters for state and parameter estimation in batch processes. J. Process Control **15**(6), 665–673 (2005)
11. Chib, S.: Calculating posterior distributions and model estimates in Markov mixture models. J. Econ. **75**, 79–97 (1996)

Electric Wheelchair Controlled-Based EMG with Backpropagation Neural Network Classifier

Arjon Turnip, Dwi Esti Kusumandari, Giovani W. G. Arson, and Daniel Setiadikarunia

Abstract Brain-controlled wheelchair is an assisting device for patients with motor disabilities controlled by brain waves. The user convenience and safety of the brain-controlled wheelchair development using EMG are focused. Patients with disabilities who are still able to move his fingers can control the brain-controlled wheelchair with a finger. This paper discusses the design and implementation of signal processing using artificial neural network for classification of motion command brain-controlled wheelchair. The signal processing is divided into three parts, namely preprocessing, feature extraction, and classification. Preprocessing stage using digital filter, FIR bandpass filter 10–500 Hz, and notch filter at 50 Hz to eliminate noise. The preprocessing proceeds at the characteristic extraction stage in the form of RMS, MAX, VAR, SD, and MAV. The value of the feature will be calculated using the artificial neural network to generate the command such as forward, turn right, turn left, or stop.

1 Introduction

The ability to move freely is the desire of every individual, especially in persons with disabilities who have limited space. Not all persons with disabilities can use their own wheelchair to travel, therefore with the help of information technology developed a smart electric wheelchair-based control using brain physiology waves. Brain physiology waves can be utilized as controllers of wheelchair movement (such

A. Turnip (✉)
Department of Electrical Engineering, Universitas Padjadjaran, Bandung, Indonesia
e-mail: jujhin@gmail.com; turnip@unpad.ac.id

D. E. Kusumandari
Technical Implementation Unit for Instrumentation Development, Indonesian Institute of Sciences, Bandung, Indonesia

G. W. G. Arson · D. Setiadikarunia
Electronics Department, Universitas Kristen Maranatha, Bandung, Indonesia

E. Joelianto et al. (eds.), *Cyber Physical, Computer and Automation System*, Advances in Intelligent Systems and Computing 1291,
https://doi.org/10.1007/978-981-33-4062-6_13

as forward, stop, turn right, or left turn) with the recording and analysis of brain biosignals [1]. To record biosignals of the brain, electroencephalogram (EEG) is used. The EEG is a device normally used to record bioelectrical signal activity through placed electrodes on the scalp [2]. In the field of medicine, the EEG is also used for diseases diagnose such as Alzheimer's and epilepsy. In addition, EEG can also be applied as a certain motor controller by utilizing bioelectrical signal extraction results. With readings obtained from the EEG, it is possible to group one's thoughts in waveform. This can then be used as information for controlling wheelchair movement by adjusting to the trained data.

In the signal processing, the limitation is often found when the subjects less concentrate or panic such that the recorded EEG data cannot be accurately used to move the wheelchair according to user orders. Therefore, an additional safety tool is needed, and one method is to add an electromyogram (EMG) as an optional controller. The EMG is capable of recording electrical signals in human muscles during contractions [3]. The finger is observed as the controlling input of wheelchair movement. The EMG is attached to the extensor digitorum muscle and the extensor policies brevis or abductor pollicis longus (because the movement of the finger gives contraction to the muscle). The EMG signal will be retrieved using RMS, MAX, VAR, SD, and MAV parameters [4]. The parameters are used as input classification using ANN. The ANN output determines the command to be executed in the brain-controlled wheelchair.

2 Methods

Brain-controlled wheelchair is a technology that combines electric wheelchairs with brain–computer interface (BCI) with the aim of facilitating wheelchair control for people with motoric disorders to move. The BCI could directly communicate between the computer and the brain by representing the results of thoughts into the control of the wheelchair movement to the desired user. This allows the user to move the wheelchair using commands from the brain's physiological signals when thinking. The recorded electrical activity called biosignal (firstly found by Hans Berger) of the brain can be performed noninvasive by the electrodes placed on the scalp surface [5]. The biosignal in the form of various electric potentials on the surface of the scalp then measured the difference of electrical potential between the two electrodes.

Electromyography is an experimental technique involving the development, recording, and analysis of myoelectric signals. Myoelectric signals are formed by physiological variations that occur in muscle fiber membranes [6]. Physiological muscles are represented by muscle fibers that can be stimulated by nerve control. Similar to electrical activity in the brain, muscles are also capable of generating electrical potentials in muscle cells. Potential is when the motor does action due to the cycle of depolarization and repolarization in muscle cells. This cycle performs a depolarization wave or electrical dipole that travels along the surface of the muscle fiber. One of the important things on taking EMG signals is the selection of good frequency sampling. At the EMG, almost all of the signal power is between 10 and 250 Hz, and it is recommended to use bandpass amplifiers at 10–500 Hz. This will result in a sampling frequency of at least 1000 Hz or even 1500 Hz to avoid loss of signal.

Bandpass filter is used to limit the frequency of the signal based on the desired frequency range. The signal frequency can be passed between the cut-off frequency of the high-pass filter and the cut-off frequency of the low-pass filter which can be adjusted, respectively. Frequencies outside the area will be muted [7]. Bandstop filter is a type of frequency selective circuit that inversely functions from a bandpass filter. Bandstop filters, known as band reject filters, pass through all the frequencies with the exception of a particular highly attenuated stop band. When the stop band has high attenuation and narrows at a certain frequency, it is commonly known as a notch filter [8].

The EMG signal is defined as a measure of electrical activity produced by muscles movements. It can be used in handling electronic devices or prosthesis. If we are able to identify the hand gesture recorded using EMG sensor with better reliability and sufficient classification rate, it could serve a good purpose for handling the devices and to provide the good quality of life to disabled subject. Artificial neural networks (ANN) are computational systems that are inspired from biological neural networks formed as generalizations of mathematical models. This system is able to recognize something that has been experienced or known, or able to do the learning process of something [9–14]. Learning on the ANN is a continuous process of adding knowledge, which will be used as a reference to recognize something similar. Information processing is carried out in neurons, and then processing proceeds from one neuron to another which is used to produce the desired output [15].

3 Design and Application

The aim of the signal processing is to identify and determine the motion of brain-controlled wheelchair, such as forward, turn right, turn left, and stop. The obtained signal is the contraction results of the index and thumb fingers of the right and left arm. This signal is recorded to produce raw data, and at preprocessing stage the signal is filtered by bandpass filter at 10–500 Hz and notch filter at 50 Hz to eliminate unnecessary signals (noise). The parameters used feature for extraction and classification are root mean square (RMS), maximum amplitude (MAX), variance (VAR), standard deviation (SD), and mean absolute value (MAV). The used classifier is ANN with feed-forward backpropagation logarithm with one hidden layer and consists of 50 nodes to determine the desired output movement.

The EMG signal recording process is performed on the surface of the forearm skin, where the position of the electrode laying is in the extensor digitorum muscle and the extensor pollicis brevis or abductor pollicis longus muscle. It is possible to observe the formed signals when moving the index and thumb fingers. Recording

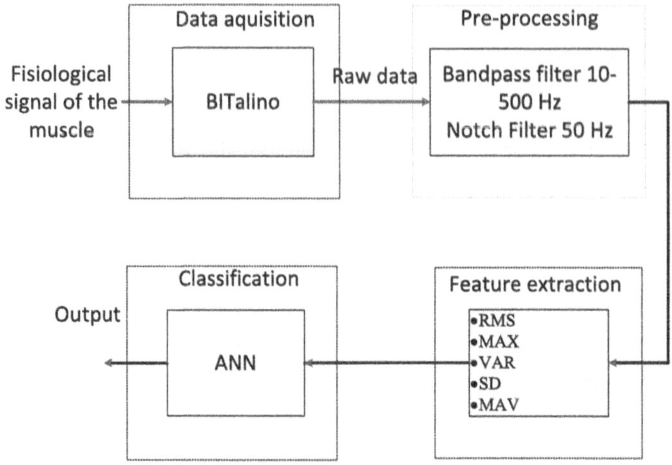

Fig. 1 Block diagram of the signal processing

experiment was conducted with ten subjects by which every finger is recorded about five times.

At this stage, the raw signal from EMG is filtered with a bandpass filter in the range of 10–500 Hz and notch filter at 50 Hz to eliminate noise that interferes with the data. The use of a bandpass amplifier at 10–500 Hz is due to the EMG signal power being in the 10–250 Hz range, otherwise a minimum sampling frequency of 1000–1500 Hz is required. While the use of notch filter aims to eliminate interference caused by electronic objects around the experimental location. After going through the preprocessing stage, the data is segmented into one second to facilitate the retrieval of features that focus on the occurrence of contractions in the muscles when the fingers are moved. Block diagram of the signal processing is shown in the Fig. 1. The features used are RMS, MAX, VAR, SD, and MAV.

In the classification using ANN, it is necessary to have data training before testing. The created ANN network structure is feed-forward backpropagation, using five inputs in the form of previous characteristic extraction results (RMS, MAX, VAR, SD, and MAV), while the hidden layer used is a hidden layer with 50 nodes, where the target for each command is made based on the fingers: 1 for the right index finger, 2 for the right thumb, 3 for the left forefinger, and 4 for the left thumb.

4 Methods

In this paper, we have worked on recognizing the combined four finger movement captured using two-channel EMG sensor. The signals have been classified using the backpropagation neural networks. Five time domain features mobility has been used to uniquely represent the EMG channel data. Tuning parameters like number of hidden layers, learning constant, and number of neighbors have been determined from the experimental results to achieve the better classification results.

On the extraction and classification test, the accuracy of finger grouping is varied. By giving a rounding limit, if the output is less than 1.5, then the value will be rounded to 1. If it is more than or equal to 1.5 and less than 2.5, the value will be rounded to 2. If the value is more than or equal to 2, 5 and less than 3.5, it is rounded to 3. If more than or equal to 3.5, it will be rounded to 4. From the classification results of the recorded data about 50 samples from the right index finger, it was obtained about 30 samples that match the desired target. These results indicate an accuracy of 60% for the turn right command. Whereas for the forward command taken from the right thumb about 50 samples with 40 samples according to the target or an accuracy of 80%. On the left hand, the index finger of 50 samples obtained 39 samples accordingly, has an accuracy of 78% for the left motion command, and the thumb of 50 samples obtained 42 samples corresponding to 84% for the stop command. The classification results of all the subject are given in Table 1. All the yellow color background indicates the success classification results in which the movement finger correctly corresponds with the intended wheelchair direction. Individually, the higher classification accuracy is obtained from subjects 1, 2, 4, and 8, and the lower classification accuracies are obtained from subjects 5 and 6.

5 Conclusions

EMG is very naturally measured when the user indicating a certain direction and the force information which will be used for the speed of wheelchair are easily extracted from EMG. The EMG signals are classified from the pre-defined motions such as rest case, forward, left, and right movements by backpropagation neural networks. The classification accuracies of 60, 80, 78, and 84% from right-hand index finger, right-hand thumb, left-hand index finger, and left-hand thumb, respectively, are obtained. The classification and evaluation results with real users show the feasibility of EMG as an input interface for powered wheelchair.

Table 1 Classification results of the right and left hand

Subject	Finger	Right Arm			Left Arm		
		Target	Output	Finger Output	Target	Output	Finger Output
Subject 1	Index	1	1.413213	Index	3	3.273178	Index
		1	1.406215	Index	3	3.447012	Thumb
		1	1.411267	Index	3	3.228697	Index
		1	1.415352	Index	3	3.256749	Index
		1	1.415755	Index	3	3.120137	Index
	Thumb	2	1.678615	Thumb	4	3.865776	Thumb
		2	1.526592	Thumb	4	3.712085	Thumb
		2	1.561857	Thumb	4	3.941523	Thumb
		2	1.571354	Thumb	4	3.569847	Thumb
		2	1.602169	Thumb	4	3.921072	Thumb
Subject 2	Index	1	1.437138	Index	3	3.108012	Index
		1	1.404255	Index	3	3.117417	Index
		1	1.40492	Index	3	3.212184	Index
		1	1.443304	Index	3	3.199339	Index
		1	1.399295	Index	3	3.10952	Index
	Thumb	2	1.484592	Index	4	3.919825	Thumb
		2	1.676701	Thumb	4	3.540536	Thumb
		2	1.594279	Thumb	4	3.179327	Index
		2	1.706609	Thumb	4	3.450967	Thumb
		2	1.647578	Thumb	4	3.861388	Thumb
Subject 3	Index	1	1.630769	Thumb	3	3.405793	Index
		1	1.497437	Index	3	3.582258	Index
		1	1.818461	Thumb	3	3.303564	Index
		1	1.323578	Index	3	3.317801	Index
		1	1.57267	Thumb	3	3.256469	Index
	Thumb	2	1.777559	Thumb	4	3.503566	Index
		2	1.286053	Index	4	3.860146	Thumb
		2	1.968623	Thumb	4	3.980085	Thumb
		2	1.79234	Thumb	4	3.979213	Thumb
		2	1.919454	Thumb	4	3.961527	Thumb
Subject 4	Index	1	1.489228	Index	3	3.027987	Index
		1	1.344879	Index	3	3.033474	Index
		1	1.642328	Thumb	3	3.044273	Index
		1	1.467193	Index	3	3.043093	Index
		1	1.51316	Thumb	3	3.034276	Index
	Thumb	2	1.521439	Thumb	4	3.560705	Thumb
		2	1.632153	Thumb	4	3.433134	Thumb
		2	1.570959	Thumb	4	3.617993	Thumb
		2	1.706893	Thumb	4	3.594354	Thumb
		2	1.4709	Index	4	3.546628	Thumb
Subject 5	Index	1	1.654721	Thumb	3	3.223673	Index
		1	1.850405	Thumb	3	3.456532	Index
		1	1.683121	Thumb	3	3.675419	Thumb
		1	1.635556	Thumb	3	3.666426	Thumb
		1	1.654226	Thumb	3	3.194399	Index
	Thumb	2	1.810893	Thumb	4	3.655683	Thumb
		2	1.845009	Thumb	4	3.580573	Thumb
		2	1.420649	Index	4	3.697056	Thumb
		2	1.667769	Thumb	4	3.67494	Thumb
		2	1.419293	Index	4	3.711803	Thumb
Subject 6	Index	1	1.478413	Index	3	3.540594	Index
		1	1.579985	Thumb	3	3.498092	Index
		1	1.412748	Index	3	3.32751	Index
		1	1.518507	Thumb	3	3.51013	Index
		1	1.453469	Index	3	3.241599	Index
	Thumb	2	1.561794	Thumb	4	3.418302	Index
		2	1.755051	Thumb	4	3.350736	Index
		2	1.74166	Thumb	4	3.485007	Index
		2	1.709429	Thumb	4	3.429505	Index
		2	1.634501	Thumb	4	3.396839	Index
Subject 7	Index	1	1.651356	Thumb	3	3.540709	Thumb
		1	1.739613	Thumb	3	3.545483	Thumb
		1	1.574152	Thumb	3	3.48599	Thumb
		1	1.584592	Thumb	3	3.526873	Thumb
		1	1.502249	Thumb	3	3.459256	Thumb
	Thumb	2	1.676496	Thumb	4	3.549493	Thumb
		2	1.702635	Thumb	4	3.540346	Thumb
		2	1.693287	Thumb	4	3.55597	Thumb
		2	1.526725	Thumb	4	3.516095	Thumb
		2	1.66638	Thumb	4	3.547409	Thumb
Subject 8	Index	1	1.072988	Index	3	3.132391	Index
		1	1.08894	Index	3	3.185395	Index
		1	1.01312	Index	3	3.23049	Index
		1	1.011465	Index	3	3.169666	Index
		1	1.01106	Index	3	3.163281	Index
	Thumb	2	1.66236	Thumb	4	3.780672	Thumb
		2	1.822357	Thumb	4	3.747669	Thumb
		2	1.790696	Thumb	4	3.579805	Index
		2	1.729365	Thumb	4	3.7198	Thumb
		2	1.649541	Thumb	4	3.665531	Thumb
Subject 9	Index	1	1.481572	Index	3	3.047259	Index
		1	1.389391	Index	3	3.042396	Index
		1	1.504959	Thumb	3	3.079641	Index
		1	1.582002	Thumb	3	3.081508	Index
		1	1.511335	Index	3	3.073757	Index
	Thumb	2	1.43979	Index	4	3.525129	Index
		2	1.470108	Index	4	3.856702	Thumb
		2	1.500207	Thumb	4	3.777942	Thumb
		2	1.534343	Thumb	4	3.766324	Thumb
		2	1.489622	Index	4	3.513551	Thumb
Subject 10	Index	1	1.152833	Index	3	3.713922	Thumb
		1	1.126949	Index	3	3.686299	Thumb
		1	1.120257	Index	3	3.298401	Index
		1	1.129499	Index	3	3.789972	Thumb
		1	1.125613	Index	3	3.449957	Thumb
	Thumb	2	1.562754	Thumb	4	3.970376	Thumb
		2	1.457157	Index	4	3.95116	Thumb
		2	1.581384	Thumb	4	3.953684	Thumb
		2	1.475838	Index	4	3.942871	Thumb
		2	1.50753	Thumb	4	3.949285	Thumb

Acknowledgements This research was supported by the Technical Implementation Unit for Instrumentation Development, Indonesian Institute of Sciences, Indonesia.

References

1. Turnip, A., et al.: EEG-based brain-controlled wheelchair with four different stimuli frequencies. Internetw. Indonesia J. **8**(1), 65–69 (2016)
2. Padfield, N., Zabalza, J., Xhao, H., Marsero, V., Ren, J.: EEG-based brain-computer interfaces using motor-imagery: techniques and challenges. Sensor **19**, 1423–1457 (2019)
3. Riillo, F., et al.: Optimization of EMG-based hand gesture recognition: supervised Versus unsupervised data preprocessing on healthy subjects and transradial amputees. Biomed. Signal Process. Control **14**(1), 117–125 (2014)
4. Nazmi, N., Rahman, M.A., Yamamoto, S.-I., Ahmad, S., Zamzuri, H., Mazlan, S.: A review of classification techniques of EMG signals during Isotonic and isometric contractions. Sensors **16**(8), 1304 (2016)
5. Carlson, T., Millan, J.R.: Brain-controlled wheelchair: a robotic architecture. IEEE Robot. Autom. **20**(1), 65–73 (2013)
6. Konrad, P.: The ABC of EMG. Noraxon U.S.A. Inc., Scottsdale (2006)
7. Zölzer, Udo. DAFX: In: Digital Audio Effects, 2nd edn. (2011). https://doi.org/10.1002/978 1119991298
8. Bandstop filter: https://www.electronics-tutorials.ws, 30 Maret (2017)
9. Turnip, A., Hong, K.-S., Jeong, M.Y.: Real-time feature extraction of P300 component using adaptive nonlinear principal component analysis. Biomed. Eng. OnLine**10**(3)
10. Turnip, A., Hong, K.-S.: Classifying mental activities from EEG-P300 signals using adaptive neural network. Int. J. Innov. Comp. Inf. Control **8**(7) (2012)
11. Philip, J.T., George, S.T.: Visual P300 mind-speller brain-computer interfaces: a walk through the recent developments with special focus on classification algorithms. Clin. EEG Neurosci. **51**(1), 19–33 (2020)
12. Turnip, A., Simbolon, A.I., Amri, M.F., Setiadi, R.H., Mulyana, E.: Backpropagation neural networks training for EEG-SSVEP classification of emotion recognition. Internetw. Indonesian J. **9**(1), 53–57 (2017)
13. Turnip, A., Amri, M.F., Suhendra, M.A., Kusumandari, D.E.: Lie detection based EEG-P300 signal classified by ANFIS method. J. Telecommun. Electron. Comput. Eng. **9**(1–5), 107–110 (2017)
14. Jiang, X., Bian, G.-B., Tian, Z.: Removal of artifacts from EEG signals: a review. Sensor **19**, 987–1005 (2019)
15. Gardner, M.W., Dorling, S.R.: Artificial neural networks (the multilayer perceptron) a review of applications in the atmospheric sciences. Atmos. Environ. **32**(14–15), 2627–2636 (1998)

Effect of Methadone on the Brain Activity in Close Eyes Condition

Arjon Turnip, Dwi Esti Kusumandari, Siti Aminah Sobana, Arifah Nur Istiqomah, Teddy Hidayat, Shelly Iskandar, Yumna Nabila, Ririn Amrina, and Putri Madona

Abstract Drug abuse in various parts of the world is increasingly widespread. Therefore, a drug addict should immediately stop and must be recovered. To overcome the symptoms of addiction, the use of methadone as a synthetic drug to replace opioid type drugs is recommended. In this paper, an experiment with rehabilitation patients to identify the effect of the drugs on the brain activity in the frontal, central, temporal, and occipital lobes is proposed. The EEG data collection is performed using 18 channel electrodes, namely central: C3, C4; frontal: Fp1, Fp2, F3, F2, F4, F7, F8; occipital: P3, Pz, P4, O1, O2; and temporal: T3, T4, T5, T6. In the brain signals record, subjects were asked to comfortably sit in a chair. The recording was done in three sessions: 5 min before drinking methadone, 10 and 60 min after drinking the methadone, respectively. To reduce background noise and artifacts removal, band-pass filter (0.5–50 Hz) and wavelet method were applied, respectively. From this experiment, it was found that a decrease in amplitude after methadone intake for average in four lobes is obtained. The result indicates that the use of methadone is highly effected on the entire brainwave activity which indicates a decrease in the level of desire to do activities.

Keywords EEG-P300 · Methadone · Visual stimuli · Frontal lobe · Brain activity

A. Turnip (✉)
Electrical Engineering, Universitas Padjadjaran, Bandung, Indonesia
e-mail: jujhin@gmail.com; turnip@unpad.ac.id

D. E. Kusumandari
Technical Implementation Unit for Instrumentation Development, Indonesian Institute of Sciences, Bandung, Indonesia

S. A. Sobana · A. N. Istiqomah · T. Hidayat · S. Iskandar
Faculty of Medicine, Padjadjaran University, Sumedang, Indonesia

Y. Nabila · R. Amrina · P. Madona
Faculty of Telecomunication Electronic, Politeknik Caltex Riau, Pekanbaru, Indonesia

E. Joelianto et al. (eds.), *Cyber Physical, Computer and Automation System*, Advances in Intelligent Systems and Computing 1291,
https://doi.org/10.1007/978-981-33-4062-6_14

1 Introduction

Drug abuse in various parts of the world is increasingly widespread. Various cases show material and non-material losses and even cause the death. Therefore, a drug addict should immediately stop and must be recovered. According to Indonesian law, narcotics addicts and victims of narcotics abuse must serve out medical rehabilitation and social rehabilitation [1]. It regulates that narcotics addicts and narcotics abuse victims who are undergoing the process of investigation, prosecution, and trial in court are supported with treatment and recovery in rehabilitation institutions (BNN Regulation 11/2014) [2]. By law, the state is responsible for recovering drug users through rehabilitation. Therefore, there should be no obstacles for rehabilitation programs, including regarding infrastructure or facilities for the recovery of drug addicts. Drug rehabilitation consists of three stages namely medical rehabilitation (detoxification), social or non-medical rehabilitation, and advanced development. Some detoxification techniques include cold turkey method where the patient is locked up in the withdrawal phase, substitution or replacement therapy where the needs of opioid or heroin addicts are replaced with other types of drugs such as methadone, or symptomatic therapy where drug administration is adjusted to the user's complaints.

The therapeutic method with an effective medical approach that still recognized today is a drug switching program to another substance called methadone therapy [3–8]. There are a variety of positive benefits that allow patients to be able to carry out their normal activities, but methadone therapy also causes side effects and the dependence that can psychologically affect the patient's quality of life [9–12]. Methadone is a therapy used for drug addicts from opioid groups such as heroin, morphine, and codeine including methamphetamine. The methadone therapy must be routinely done. Methadone is a group of opiate analgesics that can be used to treat ongoing severe pain (such as pain due to the cancer). This substance works directly in the brain by changing how the body feels and how the body responds to the pain. Methadone is also used to treat dependence on narcotic drugs (such as heroin) as an approved therapy program. It can also help prevent withdrawal symptoms due to the drug withdrawal [7–9, 12]. The success of substitution therapy such as the methadone program for drug addicts is higher than rehabilitation without drugs or detoxification. Even with this therapy, the spread of HIV can be suppressed because the use is done by drinking. Some researchers have found that methadone maintenance can significantly reduce craving symptoms except in patients with heroin dependence. Long-term consumption of heroin causes adaptive changes in the brain system that may last for a long time [13]. Verdejo et al.'s research [14] has found that methadone itself has the side effect of causing cognitive impairment. Other researchers have found that rehabilitation can effectively repair impaired cognitive function caused by buprenorphine, placebo, and methadone [15].

Electroencephalogram (EEG) is an activity that records spontaneous brain activity in the form of potential electrical signals along the scalp produced by interconnected neurons. Among the medical use of EEG, among others, for the diagnosis of diseases associated with brain and psychiatric disorders [1, 16, 2–18]. EEG is also applied to detect a person's mind patterns or mental condition. Visual observation of the EEG signal directly is very difficult given the amplitude of the EEG signal is so low and the pattern is very complex. Besides that, EEG signals are strongly influenced by various variables, including mental condition, health, activity of the patient, recording environment, electrical disturbances from other organs, external stimulation, and age of the patient. The nature of EEG signals in general is nonstationary and random so that adds complexity to the processing of EEG signals [17–20]. However, the classification of EEG signals to changes in certain variables can explain the work function of the brain and capture changes in brain activity to the relevant variable.

EEG signal in a person, generally consists of wave components which are distinguished based on their frequency region, namely: Human brain waves have a range of frequencies and amplitudes—different so that it is divided into several types of waves, namely: Delta waves (when deep asleep and without dreaming) have the frequency is less than 4 Hz with an amplitude of about 10 μV. Theta waves (occurring when light sleep or drowsiness) have frequencies between 4–8 Hz with an amplitude of around 10 μV; alpha waves (occur when relaxation or transition between conscious and unconscious states) have a frequency between 8–13 Hz with an amplitude of around 50 μV. Beta waves (in a state of thinking or in the activity) have a frequency between 13–19 Hz with an amplitude between 10–20 μV. Gamma waves (experiencing very high mental activity such as fear, very panic, appearing in public) have a frequency between 19–100 Hz [21, 22]. Therefore, the representation of EEG signals into the frequency domain is mostly done in research related to EEG signal analysis. In this study, the use of EEG signals to observe the effect of methadone administration on changes in brain activity in the central, frontal, parietal, occipital, and temporal parts is proposed. So far, methadone experiments and observations of their effects on brain activity using brain waves from EEG signals are still rarely done.

2 Method

Experiments were carried out in a room that was conditioned away from noise and provided comfort for the subject (Fig. 1). Before conducting an EEG signal recording session, the subjects first directed interviews with the medical team, filled out information of concern, and follow the urine tests. Then, the subject is attached to an instrument in the form of an electro-cap on the head and also tied a belt to the chest of the subject so that the electro-cap does not shift. Subjects were briefed regarding the experimental scenario. The EEG signals are recorded through 19 channel electrodes, including Fp1, Fp2, F7, F3, F2, F4, F8, T3, C3, Cz, C4, T4, T5, P3, P2, P4,T6, O1, and O2. The reference in this experiment uses electrodes mounted on the ear, the A1 and A2 channel electrodes, which are A1 for the left ear and A2 for the right ear. When installing electrodes, a gel is applied to each EEG sensor to increase conductivity while maintaining impedance between the scalp and electrodes below 5 kΩ. The electrode mounting position, the electrolyte liquid used in the form of

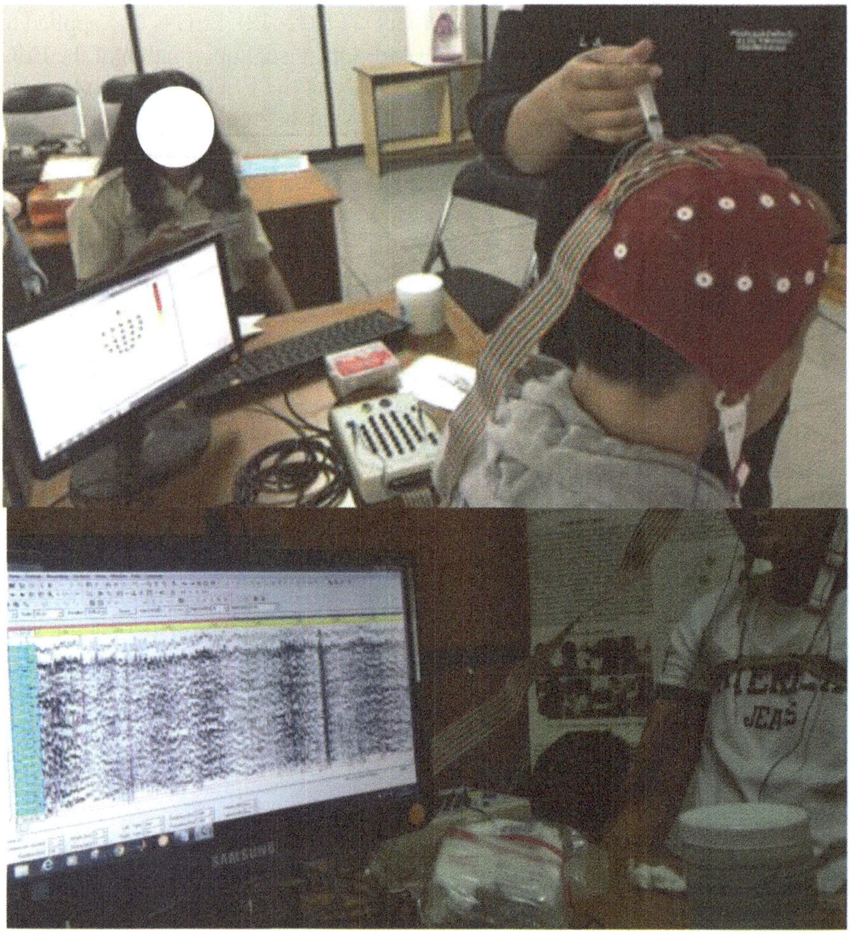

Fig. 1 Experiment design

an electrogel, and the impedance shown in Fig. 2. The electrode impedance can be monitored in WinEEG software before the signal recording process is started. The dark color on the electrode indicator indicates a higher impedance, while the bright color indicates a low impedance levels.

In addition to fix the electrode impedance, another thing to consider is setting the recording process in the WinEEG system. Settings include the sampling frequency used which is 500 Hz, the list of channels to be used and their references, and interconnection with PCs/laptops to display the stimuli used when recording EEG signals. During the recording process, subject was asked to sit relaxed while closing his eyes. Experimental time allocated for each trial is 1 h and 10 min. The time is divided into three sessions, namely 5 min before, 10 min after, and 60 min after consuming methadone. After the recording session is finished, the recording of the

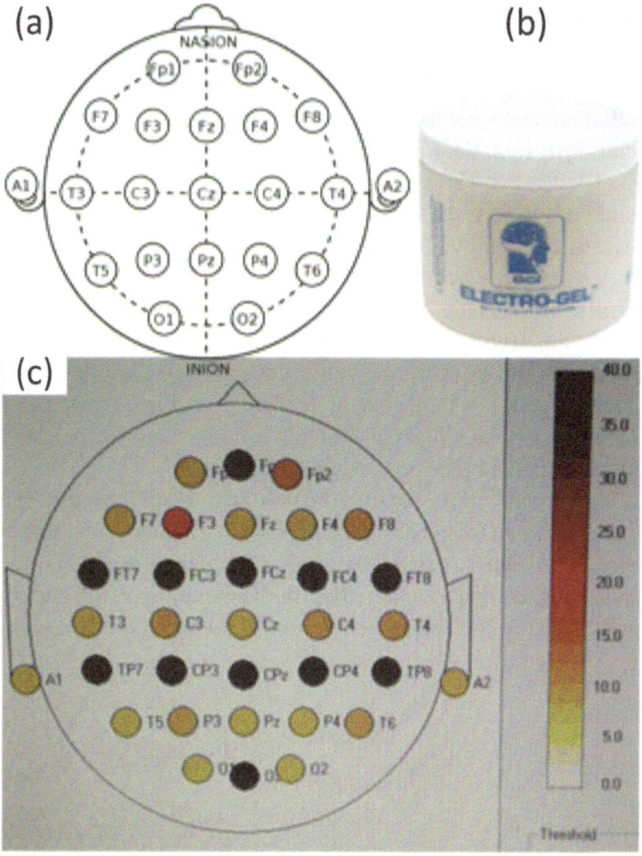

Fig. 2 **a** Electrode position, **b** Electrogel for conductivity, **c** Electrode conductivity with around 5 kΩ impedance

EEG signal is exported into a file with EEG format, which the file can later be processed using a signal processor. Each subject was supported with consumption and transportation about Rp 150,000. Ethical clearance is a written statement provided by the Research Ethics Commission for research involving living things which states that a research proposal is feasible to be carried out after fulfilling certain requirements. Prior to the experiment, this research was completed with ethical clearance from the Health Research Ethics Committee—Faculty of Medicine, Universitas Padjadjaran and each subject was asked to fill out inform of concern.

3 Signal Processing

EEG raw data is processed with Cz references using 18 channels from 8 subjects. The 18 channels that are used are grouped according to the brain lobes, namely central: C3, C4; frontal. Fp1, Fp2, F3, F2, F4, F7, F8; parietal occipital: P3, Pz, P4, O1, O2; temporal: T3, T4, T5, T6. The average amplitude (after extraction) of each channel group is calculated. Before the data is processed montage reference is changed to the middle part of the brain with the Cz channel. Recording of each subject is done for ± 5 min per session. Data is taken from 10 to 130 s because data processing will be more effective if taken 2 min of data that is clean and free of artifacts. Data recording before the 10th second is cut because the initial 10 s are considered to be still corrupted by noise where the subject is still adjusting to the experimental conditions. From Fig. 3, raw data generated, clearly visible on the EEG signal there are still many artifacts which make it difficult in understanding the character of the signal, therefore, the next processes are needed.

To reduce background noise, the filtering process for EEG raw data is carried out. Bandpass filter is a circuit that is designed to pass the frequency within certain limits and reject other frequencies outside the desired frequency. And bandpass filter is a combination of highpass and lowpass filter. In the experiment, the cut off frequency used is 0.5 and 50 Hz. As for feature extraction, the wavelet method with the symlet

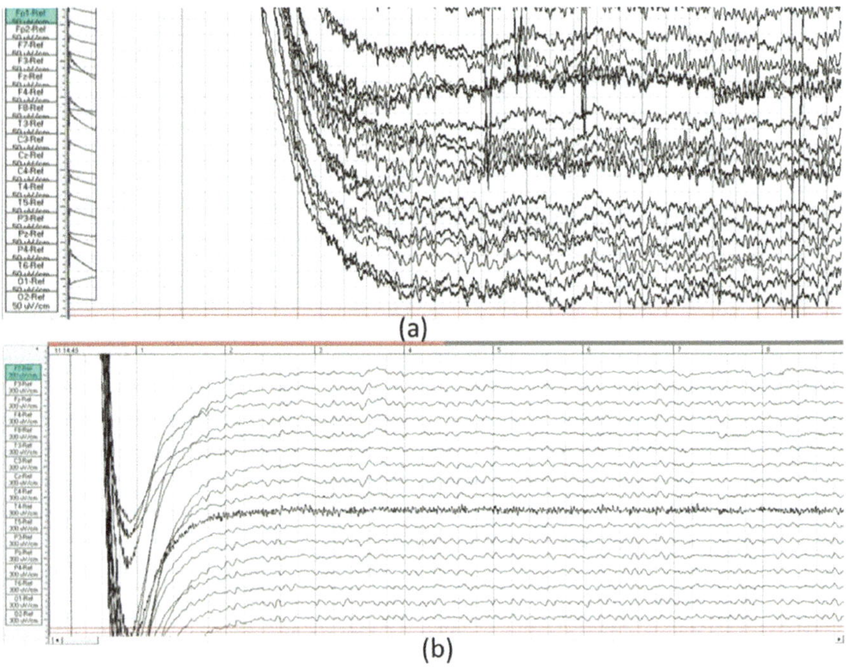

Fig. 3 Raw data of EEG in relax and close eye condition: **a** before and **b** after Methadone intake

model and the five-level decomposition process is used. Wavelet transform is a signal processing method by resembling signal analysis using Fourier transforms, namely by breaking the signal to be analyzed into several parts. The difference, if the Fourier transform signal is broken down into signal sinusoids with different frequencies, then the wavelet transformation of the analyzed signal is broken down into a number of signals resulting from shifting and scaling of a small signal called a wavelet. The wavelet transform method in Eq. (1) [23, 24], mainly used to identify true components and remove noise from the raw data:

$$W_f(j,k) = \int_{-\infty}^{+\infty} f(t)\psi_{j,k} * (t)\mathrm{d}t, \tag{1}$$

where, $\psi_{j,k}(t) = 2^{-j/2}\psi(2^{-j}t - k$, where $\psi(t)$ is the mother wavelet, $f(t)$ is the series analyzed, and t indicates the time; integer j indicates the decomposition level, and k indicates the time translation factor, and $\psi * (t)$ is the complex conjugate. The first step of the wavelet application starts from the original signal then the coefficients set is approximated on each level. In each step except the first one, only the approximated coefficients are analyzed. The wavelet used must meet the regularity of order N condition in Eq. (2):

$$\int_{-\infty}^{+\infty} t^k\psi(t)\mathrm{d}t = 0, \quad k = 1, \cdots, N - 1 \tag{2}$$

Under the level of j, the original signals can be reconstructed using Eq. (3).

$$f_j(t) = \sum_k W_f(j,k)\psi^*(2^{-j}t - k) \tag{3}$$

By increasing the decomposition level j, the detailed information of signals at larger temporal scales would obtained. The more contribute information we have, the better performance of the model is achieved. However, more input could reduce the computing efficiency and decrease the stability of the model. Therefore, it is important to select an appropriate decomposition level for wavelet modeling [23].

4 Results and Discussion

Drugs provide a dominant effect on the functioning of the four brain lobe: frontal, parietal, temporal, and occipital lobes. These effects can be observed if brain activities are recorded and processed. In the experiment of brain activity record, a group of subject is asked to sit in relax while closing their eyes. The brain wave was recorded about three times: before, 10 min, and 60 min after taking methadone. Assumption

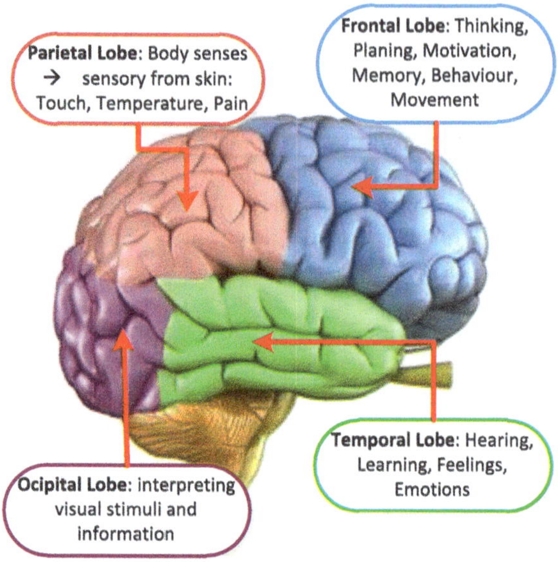

Fig. 4 Brain lobe effected by methadone

that subjects who are following the rules of experiment will fill craving in the first record, starts to comfort after 10 min, and feel comforts after 60 min of consuming methadone. The differences in the amplitude of the extracted EEG before and 1 h after the subject consume the methadone in four region of brain is observed. Subjects who have not been given methadone have a higher level of interest in methadone (craving), consequently, the amplitude after consuming methadone must be lower. The decrease in amplitude value was also supported by the influence of methadone which tends to make the subject sleepy where theta waves increase and beta waves decrease.

In this research, the cerebral cortex of the human brain was divided into four lobes (see Fig. 4): the frontal, parietal, occipital, and temporal lobes. The four lobes of the brain have their respective functions in the body ranging from reasoning to auditory perception. The frontal lobe is usually associated with the ability for motivation, concentration, movement, cognition, and language understanding. Damage to the frontal brain is usually characterized by changes in sexual habits, socializing, and attention also has problems with risk taking. The parietal lobe is located behind the frontal lobe. The parietal lobe plays a role in interpreting touch, body movements, pain sensations, and ability to count. Injury or damage to the parietal lobe can cause a person to lose sensation (numbness or tingling) on the opposite side of the body. This section has an important role in interpreting messages from other parts of the brain. The temporal lobes are located on both sides of the head which are parallel to the ear. This part of the cerebrum is responsible for auditory, memory, and emotional functions. Damage to the temporal lobe can cause problems with memory, speech perception, and language skills. The occipital lobe is located in the back of the brain. This part of the cerebrum is useful to help us recognize objects through the sense of sight and understand the meaning of written words. Damage to these lobes can cause problems in the form of difficulty recognizing objects, inability to identify colors, hallucinations, and difficulty understanding words [23].

The danger of drug abuse for health is of course we 'have often heard. Of the many adverse effects caused, drug use can affect brain performance. Because it is a vital organ as well as the body's control center, the brain affected by drugs will affect all bodily functions. Some of the drug consumption effects on brain function are manipulating mood and behavior, triggering the brain to work extra (so that the user feels refreshed, enthusiasm, and self-confidence increase) blood pressure increases and inhibits the work of the brain called depression. There are also drugs that cause delusions, or what are often called hallucinogens. Narcotics abuse has an influence on the work of the nervous system including: Sensory nerve disorders (central and occipital lobes) that cause numbness and blurred vision that can cause blindness; autonomic nerve disorders (frontal lobes) that cause undesired movements through motor movements. Motor nerve disorders (central and frontal lobes) that cause loss of coordination with the motor system. Vegetative nerve disorders (frontal, temporal, and central lobes) cause language to come out of consciousness and cause fear and lack of confidence.

The average amplitude of each lobes before and after consuming methadone is given in Table 1. If referring to the results of previous studies, the use of drugs will automatically affect the performance of the brain in each lobes such as disturbing vision for occipital, movement and language for central and frontal, emotions for temporal. In theory, if someone who is being carved is given methadone then the subject should feel more comfortable after one hour. Based on the experimental results in Table 1, except for subjects 5 and 6, the average amplitude in each lobes decreased. Individually, the subtexts 2, 3, 4, 7, and 8 have an increase in amplitude, respectively, in the central, occipital, frontal and temporal, occipital, central, and frontal.

When compared with subjects 5 and 6, the increase was not so significant. Subjects 5 and 6 have lobes with increased amplitude after consuming methadone. Based on the history of substance use, the two subjects used almost the same type of drug and most compared to other subjects. Having almost the same age with a slight mental disorder and the value of impulsivity is relatively high. They also still took high doses of methadone despite having undergone rehabilitation for a long time (subject 5 had been rehap for 11 years). Subjects 5 and 6 also consumed the substance benzodiazepines during the experiment. Based on the demographic conditions of the two subjects, it can be understood that the increase in the value of brain activity amplitude in each lobes is a result of impaired brain function in the relevant part. The highest increase was seen in subject 6, which was around 6.5 times from the time of craving. While the increase in subject 5 is only about 2.5 times the condition while craving. Based on demographic conditions, subject 6 still consumed methadone at very high or maximum doses during the experiment. It is suspected that subject 6 did not participate in rehabilitation properly and routinely. While subject 5 had a significant reduction in the maximum dose. However, due to the age of the use of the drug that has been quite long which is 11 years, it is most likely that many brain

Table 1 Amplitude of brain activity in the lobe of central, frontal, occipital, and temporal: before, 10 min, one hour of methadone intake

Amplitude (microvolts)				
S	Lobes	Before	10 m	1 h
1	Central	28.29	20.82	10.01
	Frontal	119.80	63.73	8.34
	Occipital	48.23	37.78	15.40
	Temporal	69.81	63.78	20.52
2	Central	7.80	5.82	18.12
	Frontal	10.08	12.06	9.10
	Occipital	9.85	14.40	3.27
	Temporal	10.55	17.04	2.71
3	Central	216.88	131.25	55.29
	Frontal	84.02	61.15	33.26
	Occipital	89.55	105.08	124.45
	Temporal	36.77	31.04	11.39
4	Central	42.69	28.47	32.42
	Frontal	26.47	28.07	29.34
	Occipital	34.85	17.94	32.73
	Temporal	8.70	8.46	9.34
5	Central	14.46	28.87	49.23
	Frontal	15.17	21.83	50.79
	Occipital	20.55	59.43	25.65
	Temporal	4.58	7.32	18.28
6	Central	24.49	20.47	120.79
	Frontal	24.50	22.86	122.04
	Occipital	13.93	22.32	197.58
	Temporal	16.74	5.6	92.36
7	Central	40.58	11.69	5.498
	Frontal	38.51	11.85	8.81
	Occipital	1.25	6.76	5.46
	Temporal	16.65	3.37	3.49
8	Central	27.05	28.41	33.38
	Frontal	23.26	23.49	31.73
	Occipital	16.40	14.85	16.89
	Temporal	7.93	9.83	7.53

nerve tissue has been damaged so that even though methadone has been given, brain function cannot return to normal.

Subjects 4 and 8 both have two lobes whose brain activity amplitude values do not decrease. Both subjects both had the second most substance use history than any other subject and the dosage was almost the same as the maximum dose. If seen as a whole, the change in the value of the amplitude of brain activity after consuming methadone is very closely related to the history of substance use and the reduction in the dose of methadone. When compared with the subjects 2 and 3, subject 7 experienced a significant increase in amplitude in the occipital region which is almost 4 times the craving condition. Based on urine test results, subject 7 was indicated using benzodiazepines and methamine during the experiment. These conditions are sufficient to state the reason for the increase in amplitude associated.

5 Conclusion

Methadone intake by the drug rehabilitation patients causes a decrease in the brain's impulsivity which indicates a decrease in the level of desire for drugs. The main results of the present analysis indicated that the subjects have a smaller amplitude of brain activity after consuming methadone. This study revealed that drug patients have abnormalities of brain activity in central, frontal, occipital, and temporal lobes, which may reflect deficits in cognitive function.

Acknowledgements This research was supported by Technical Implementation Unit for Instrumentation Development, Indonesian Institute of Sciences and funded by RISTEKDIKTI by INSINAS 2019, Indonesia.

References

1. *Undang-Undang Republik Indonesia Tentang Psikotropika (5)* (Law of the Republic of Indonesia Concerning Psychotropics**5**, (1997). https://hukum.unsrat.ac.id/uu/uu_5_97.htm. Last Accessed
2. BNN (2007) *Survei Nasional Penyalahgunaan dan Peredaran Gelap Narkoba* (National Survey of Narcotics Abuse and Illicit Drugs) (2003). Puslitbang & Info LakharBNN Hompage. https://www.bnn.go.id. Last Accesed 11 2013
3. Wang, G.Y., Kydd, R., Wouldes, T.A., Jensen, M., Russell, B.R.: Changes in resting EEG following methadone treatment in opiate addicts. Clin. Neurophysiol. **126**(5), 943–950 (2013)
4. Yang, L., et. al.: The effects of methadone maintenance treatment on heroin addicts with response inhibition function impairments: evidence from event-related potentials. J. Food Drug Anal. **23**, 260–266 (2015)
5. Turnip, A., et al.: The increase of brain activity in frontal lobe after one hour methadone intake. In: International Conference on Sustainable Engineering and Creative Computing (ICSECC) (2019)
6. Turnip, A., Kusumandari, D.E., Hidayat, T., Hidayat, T.: Brain mapping of low and high implusivity based P300 signals. J. Phys. Conf. Ser. **1007** (2018)

7. Hu, B., et al.: Effective brain network analysis with resting-state EEG data: a comparison between heroin abstinent and non-addicted subjects. J. Neural Eng. **14**(4), (2017)
8. Wang, G.Y., Kydd, R., Wouldes, T.A., Jensen, M., Russell, B.R.: Changes in resting EEG following methadone treatment in opiate addicts. Clin Neurophysiol. **126**(5), 943–950 (2015)
9. Yang, L., et. al.:The effects of methadone maintenance treatment on heroin addicts with response inhibition function impairments: evidence from event-related potentials. J. Food Drug Anal. **23**, 260–266 (2015)
10. Maeyer, J.D., Vanderplasschen, W., Camfield, L., Vanheule, S., Sabbe, B., Broekaert, E.: A good quality of life under the influence of methadone: a qualitative study among opiate-dependent individuals. Int. J. Nurs. Stud. **48**, 1244–1257 (2011)
11. Lin, C.Y., Chang, K.C., Wang, J.D., et al.: Quality of life and its determinants for heroin addicts receiving a methadone maintenance program: Comparison with matched referents from the general population. J. Formos. Med. Assoc **115**(9), 714–727 (2016)
12. Malik, E., Adelson, M., Sason, A., Schreiber, S., Peles, E.: Outcome of patients with high depressive symptoms on admission to methadone maintenance treatment. J. Dual Diagn.1–10 (2019)
13. Li, Q., Wang, Y., Li, W., Yang, W.C., Zhu, J., Chang, H.F., Wu, N., Zheng, Y., Wang, W.: fMRI study on craving and brain activity in response to heroin-related cues in patients with methadone maintenance treatment. J. Pract. Radiol. **28**, 1–5 (2012)
14. Verdejo, A., Toribio, I., Orozco, C., Puente, K.L., Perez-Garcı´a, M.: Neuropsychological functioning in methadone maintenance patients versus abstinent heroin abusers. Drug Alcohol Depen. **78**, 283–8 (2005)
15. Attou, A., Figiel, C., Timsit-Berthier, M.: Opioid addiction: P300 assessment in treatment by methadone substitution. Neurophysiol. Clin. **31**(171), (2001)
16. Pastor, A., Conn, J., O'Brien, C.L., Teng, J., Loh, M., Collins, L., MacIsaac, R.J., Bonomo, Y.: Clinicians feel comfortable discussing alcohol but not illicit drug use with young adults with type 1 diabetes: A survey of clinicians, Diabet Med. (2019)
17. Turnip, A. et al.: Detection of drug effects on brain activity using EEG-P300 with similar stimuli. In: IOP Conference Series: Materials Science and Engineering, vol. 220 (2018).
18. Hu, B., et al.: Effective brain network analysis with resting-state EEG data: a comparison between heroin abstinent and non-addicted subjects. J. Neural Eng. **14**(4) (2017)
19. Iskandar, S., Kusumandari, D.E., Turnip, A.: Artifact removal with independent component analysis for 2D brain mapping of drugs user before and after taking methadone. Internet Working Indonesia J **11**(1), 29–33 (2019)
20. Turnip, A., Kusumandari, D. E., Pamungkas, D. S.: Drug abuse identification based EEG-P300 amplitude and, latency with fuzzy logic classifier. In: International Conference on Applied Engineering (ICAE) (2018)
21. Motlagh, F. Ibrahim, F., Rashid, R., Shafiabady, N., Seghatoleslam, T., Habil, H.: Acute effects of methadone on EEG power spectrum and event-related potentials among heroin dependents. Psychopharmacology **235**(11), 3273–3288 (2018)
22. Zhang, X., Yao, L., Kanhere, S.S., Liu, Y., Gu, T., Chen, K.: MindID: person identification from brain waves through A, ention-based recurrent neural network. ACM J. Comput. Cult. Herit. **9**(4), 39 (2017)
23. Yang, M., Sang, Y.-F., Liu, C., Wang, Z.: Discussion on the choice of decomposition level for wavelet based hydrological time series modeling. Water **8**, 197–208 (2016)
24. Jawabri, K.H., Sharma, S.: Cerebral cortex functions. Stat Pearls Publishing, Physiology, Treasure Island (2019)

Estimating Corn Weight Using Mixed Model with Linear Covariance Function Matrix

Sandy Vantika, Udjianna S. Pasaribu, Sapto W. Indratno, and Adi Pancoro

Abstract The mixed model was proposed to estimate plant breeding value. A total of 500 genotyped individual corn (Zea mays) data were taken randomly from the population which consisted of 528 genotyped individual corn data. These data were used as training data while the rest which consisted of 28 genotyped individual corn data were used as test data. We used two models which involved genomic relationship matrix (GM) and linear covariance function matrix (LCFM) to estimate phenotype values (corn weight) of 28 individual corns in test data based on their genotyped data. The correlation between estimated corn weight and true corn weight was computed for each model. We repeated the process for 100 different training and test data. The 95% confidence interval for squared correlation between estimated corn weight and true corn weight from model using LCFM was 0.162–0.209. It was better than GM model which was 0.156–0.203.

Keywords Mixed model · Training data · Linear covariance function matrix

1 Introduction

Genetic progress by selection and mating is based on prediction of the ability of the parents to breed the most efficient descendants [1]. This process of prediction is called genetic evaluation or prediction [2]. For the last century, genetic evaluation in plants and livestock has been based on the use of phenotypes at the traits of interest, together with pedigree [3]. In most cases, these evaluations ignore the physical base of heredity, i.e., DNA, and use a simplified conceptual representation of the transmission of genetic information from parents to offspring; namely, each parent passes on

S. Vantika (✉) · U. S. Pasaribu · S. W. Indratno
Faculty of Mathematics and Natural Sciences, Institut Teknologi Bandung, Bandung, Indonesia
e-mail: sandy.vantika@math.itb.ac.id

A. Pancoro
School of Life Sciences and Technology, Institut Teknologi Bandung, Bandung, Indonesia

E. Joelianto et al. (eds.), *Cyber Physical, Computer and Automation System*, Advances in Intelligent Systems and Computing 1291,
https://doi.org/10.1007/978-981-33-4062-6_15

average half its genetic constitution, associated with an unknown sampling known as Mendelian sampling [1].

We can take DNA data from an individual plant since the plant is aged early [4]. This data is known as genotype data [5]. In the genotype data, there is an event called single nucleotide polymorphism (SNP) [6]. This SNP information can be used to estimate the phenotype of the individual plant [7]. Phenotype is a visible trait of an individual. In plants, the phenotype can be a quantitative and qualitative traits [8]. Quantitative traits are such as plant height and fruit weight [9]. Qualitative traits are such as fruit taste and flower color. After obtained the estimation results, the selection process is done. This selection eliminates individuals who are incompatible with certain criteria and retain individuals who fit the criteria to be cultivated. This can save costs and time because we do not need to maintain the plants until adulthood to know the value of their phenotype later.

Recent technical developments allow stepping further into biology and peeking at the genome in the form of single nucleotide polymorphisms, known as SNP markers. Several statistical models for genomic predictions using genome-wide SNP markers have been proposed. One of the most popularly used models is genomic BLUP (GBLUP), which is a linear mixed model incorporating a marker-based genomic relationship matrix (G-matrix), because it is in the same form as a simple traditional BLUP model and has a low computational requirement [10]. The approach is based on the simultaneous estimation of allele substitution effects (ASE) for each of the markers using linear models applied to phenotypes or estimated breeding values (EBV) available on genotyped individuals comprising a training population, the determination of the accuracy of the derived prediction equations in an independent validation population and application of the prediction equations to generate genomic estimated breeding values (GEBV) in selected candidates within an implementation population. The term training population arises from the idea that some form of model is "trained" on genotypes and phenotypes to produce estimates of ASE and GEBV [11]. The purpose of the validation step is to use phenotypes available on an independent set of genotyped individuals to those used in the training population to produce an estimate of the accuracy of the GEBV that will be generated for the selected candidates. Consequently, the individuals sampled to form the validation population should be representative of the selected candidates in the sense that the accuracies of GEBV produced for the validation population should reflect the accuracies of GEBV produced for the selected candidates in the implementation population.

Association studies are regressions of phenotypes onto every single locus genotypes. The following mixed model was fitted [12]:

$$y = X\beta + Zv + \varepsilon \tag{1}$$

where y $(n \times 1)$ is the phenotype, n is number of individuals, X $(n \times p)$ and Z $(n \times q)$ are design matrices for fixed and random effects, respectively, p is number of fixed effects, q is number of random effects; $\beta(p \times 1)$ is a vector of fixed effects, $v(q \times 1)$ is additive genetic random effects, and $\varepsilon(n \times 1)$ is a vector of residuals. Random effects ν and ε had assumed distributions $qMVN\left(0, \sigma_\nu^2 I\right)$ and $nMVN\left(0, \sigma_\varepsilon^2 I\right)$, where σ_ν^2 is the additive genetic variance, σ_ε^2 is the residual variance, and I is an identity matrix. The model included fixed effects for soil condition. The effect of a SNP was fitted as a random effect and was coded 0, 1, or 2 corresponding to the number of minor allele. Vector $\alpha = Z\nu$ is defined as breeding value and had assumed $nMVN(0, G)$ where

$$G = \frac{ZZ^T}{2\sum p_j(1 - p_j)} = \frac{1}{2\sum p_j(1 - p_j)} \begin{pmatrix} z_1 \cdot z_1 & z_1 \cdot z_2 & \cdots & z_1 \cdot z_n \\ z_2 \cdot z_1 & z_2 \cdot z_2 & \cdots & z_2 \cdot z_n \\ \vdots & \vdots & \ddots & \vdots \\ z_n \cdot z_1 & z_n \cdot z_2 & \cdots & z_n \cdot z_n \end{pmatrix} \tag{2}$$

where p_j is relative frequency of minor allele for loci jth and $z_i \cdot z_j$ is dot product between row ith and row jth in matrix Z. Matrix G is also known as genomic relationship matrix. Dot product $z_i \cdot z_j$ can be seen as one kind of covariance function $k(z_i, z_j)$.

There are some novelties in this paper. Genomic relationship matrix G would be built using more general covariance function than Vanraden did [13]. Those covariance functions are linear. We also proposed a new method to estimate phenotype value in breeding material without determining estimated additive genetic random effects $\hat{\nu}$.

This paper is initially begun with introduction which consists of genomic selection background, recent technical developments and model in genomic selection, and short description of some novelties. Second, we proposed a new method to estimate phenotype value in breeding material without determining estimated additive genetic random effects $\hat{\nu}$ in the method. Third, we described the data and present a simulation in data and simulation procedure. Fourth, we explained the results in simulation results. Last, we concluded the results in conclusion.

2 The Method

Let M be the matrix that specifies which marker alleles each individual inherited. Dimensions of M are n (the number of individuals) by q (the number of loci). Elements of M are set to 0, 1, and 2 for the homozygote (*AA*), heterozygote (*Aa*, *aA*), and other homozygote (*aa*), respectively. Let the relative frequency of the minor allele at locus j be p_j, and let P contain relative allele frequencies multiplied by 2, such that column j of P is $2p_j$. Subtraction of P from M gives Z [13].

Matrix G would be built using covariance function matrix K where $k_{ij} = k\left(z_i, z_j|\theta\right)$ is conditional covariance function of z_i and z_j given parameter θ. Matrix

K is:

$$K = \begin{pmatrix} k(z_1, z_1|\theta) & k(z_1, z_2|\theta) & \cdots & k(z_1, z_n|\theta) \\ k(z_2, z_1|\theta) & k(z_2, z_2|\theta) & \cdots & k(z_2, z_n|\theta) \\ \vdots & \vdots & \ddots & \vdots \\ k(z_n, z_1|\theta) & k(z_n, z_2|\theta) & \cdots & k(z_n, z_n|\theta) \end{pmatrix} \quad (3)$$

In Eq. (1), $X\beta$ on the right side moved to the left side becoming

$$y - X\beta = Zv + \varepsilon \quad (4)$$

It is obtained $y - X\beta = Zv + \varepsilon = \delta$ which is the difference between $\boldsymbol{y}$ and its mean such that δ has distribution $nMVN(\mathbf{0}, K' = K + \sigma_\varepsilon^2 I)$. Likelihood function for θ can be written as:

$$g(\theta) = \frac{1}{(2\pi)^{n/2}|K'|^{1/2}} e^{-\frac{1}{2}\delta^T (K')^{-1}\delta} \quad (5)$$

where $\left|K'\right|$ is the determinant of covariance matrix K'. For estimating θ, it can be done by maximizing the log-likelihood function as follows.

$$\log(g(\theta)) = -\frac{1}{2}\delta^T (K')^{-1}\delta - \frac{n}{2}\log(2\pi) - \frac{1}{2}\log\left|K'\right| \quad (6)$$

Let $\delta^* = y - X^*\beta$ is δ for some new individuals. Using conditional probability definition, the conditional probability can be written as:

$$P\left(\delta^*|\delta\right) = \frac{P(\delta^*, \delta)}{P(\delta)} \quad (7)$$

The joint distribution of δ^* and δ is $(n + m)MVN(0, \Lambda)$ where

$$\Lambda = \begin{pmatrix} K'_\delta & K_{\delta,\delta^*} \\ K_{\delta^*,\delta} & K_{\delta^*} \end{pmatrix} \quad (8)$$

and K'_δ is K' for δ of training individuals, K_{δ,δ^*} is K between δ of training individuals and δ^* of new individuals and so on. From this joint distribution, we would obtain conditional probability distribution of δ^* given δ which is $nMVN(K_{\delta^*,\delta}(K'_\delta)^{-1}\delta, K_{\delta^*} - K_{\delta^*,\delta}(K'_\delta)^{-1}K_{\delta,\delta^*}$. Since $\delta = y - X\beta$, it follows

$$E\left[\delta^*|\delta\right] = E\left[y^* - X^*\beta|\delta\right] = E\left[y^*|\delta\right] - X^*\beta \quad (9)$$

where y^* and X^* are y and X for new individuals. Since $E[\delta^*|\delta] = K_{\delta^*,\delta}(K'_\delta)^{-1}\delta$, it is obtained

$$E\left[y^*|\delta\right] = X^*\beta + K_{\delta^*,\delta}(K'_\delta)^{-1}(y - X\beta) \tag{10}$$

Using Eq. (10), the estimation of new individual's phenotype could be obtained using $K_{\delta^*,\delta}$ and K'_δ which is the novelty of this paper.

3 Data and Simulation Procedure

Phenotypic data are available for 1 trait which is maize grain yield. Data are available for 100 SNP markers and 264 lines. There are two datasets with the same lines, placed in severe drought conditions and well-watered. Hence, there are 528 individuals. Genotypes are coded into 0, 1, and 2 (based on minor allele frequency for each SNP). Data were made public by Crossa et al. [14]. The statistic summary and box plot for the data are presented in Table 1 and Fig. 1.

Training population consisted of 500 individuals which were taken randomly from 528 individuals and the rest consisted of 28 individuals were used as test population. Vector y, matrix X, matrix Z, and matrix G in Eq. (2) were determined from the training population. Let b be $\hat{\beta}$, estimation of vector of fixed effects β. Vector $\boldsymbol{b}$ was computed by least squared method as:

$$b = \left(X^T X\right)^{-1} X^T y \tag{11}$$

Vector δ was also computed as:

$$\delta = y - Xb \tag{12}$$

Table 1 Statistic summary for maize grain yield data

Statistic	Dry field	Well-watered field
Number of individual plant	264	264
Minimum	0.012	0.874
Maximum	4.927	5.938
Lower quartile	1.326	2.207
Middle quartile	2.028	2.867
Upper quartile	2.727	3.581
Sum	553.753	772.254
Average	2.098	2.925
Variance	0.999	0.999
Standard deviation	0.999	0.999
Skewness	0.371	0.401
Kurtosis	−0.209	−0.005

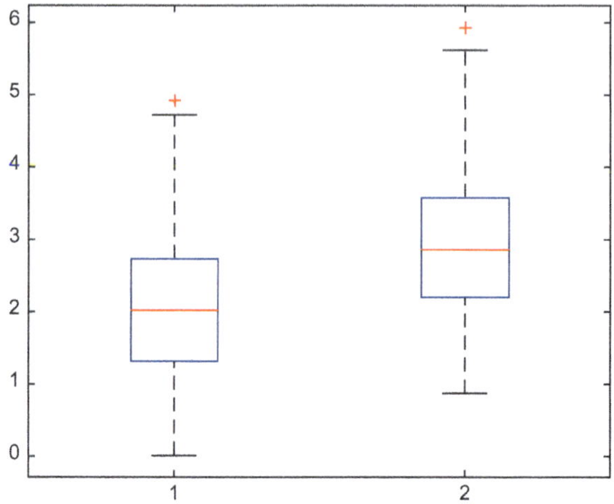

Fig. 1 Box plot of maize grain yield (corn weight) from 264 corn plants on: (1) dry field, (2) well-watered field

In the simulation, we used CLFM, K, and the original matrix G itself (GM). The covariance function was linear, $k(i, j) = \sigma_\nu^2 (z_i . z_j)$. We estimated the parameter of covariance function and σ_ε^2 using maximum likelihood method by maximizing the log-likelihood function as in Eq. (6). We evaluated parameter σ_ε^2 and σ_ν^2 from 0.1 to 1 with step 0.1. After we have estimated those parameters, the estimated CLFM was obtained. This matrix is K_δ. We have also G_δ, the original matrix $\boldsymbol{G}$ for δ of training individuals. Next, we determined $K_{\delta^*,\delta}$ using the estimated parameters. We also determined $G_{\delta^*,\delta}$, G between δ of training individuals and δ^* of new individuals. Using Eq. (10), estimated phenotype (maize grain yield), y^*, was computed for two models (CLFM, K, and GM, G).

4 Simulation Results

About 100 runs were done for two models simultaneously. Correlation between estimated phenotype (maize grain yield) and true phenotype (maize grain yield) was computed for each run. The box plots for correlations from two models obtained by 100 runs were presented in Fig. 2.

Distribution of correlation involving GM is almost similar with distribution of correlation involving LCFM.

Considering the correlation values obtained from 100 runs as a random variable, there were tested whether the correlation values involving two covariance function matrices (GM and LCFM) follow normal distribution. If the correlation values are normally distributed, then the confidence interval for the mean values of the correlation can be determined.

Normality test used in this case was chi squared test. Of the two covariance function matrices, no correlation value follows normal distribution.

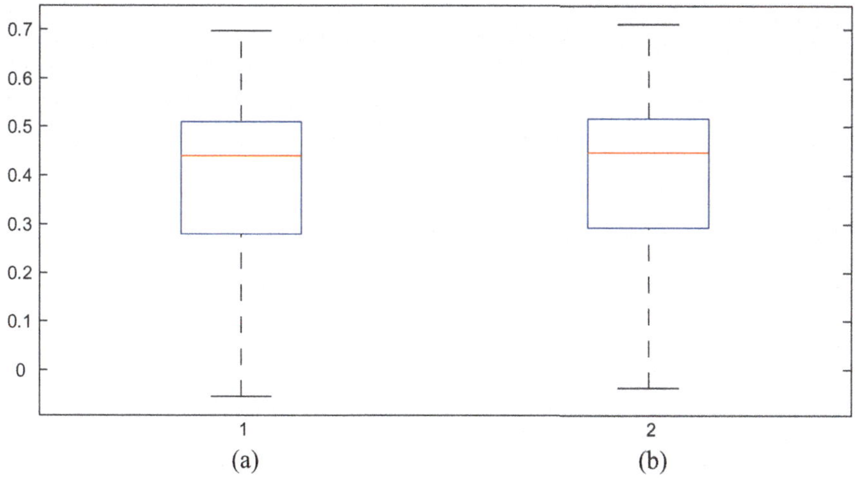

Fig. 2 Box plots of correlations: **a** GM and **b** LCFM

To compare the skewness of each correlation value distribution for both models, the histogram of correlation value for each model is presented in Fig. 3.

The skewness of correlation value distribution for model using GM was −0.6134 while for model using LCFM was −0.5767. It can be said that the skewness of correlation value distribution for both models was not closer to zero. It means that their skewness was not closer to the skewness of normal distribution.

Correlation values involving GM and LCFM were transformed to make them more symmetric. Based on the Tukey transformation, a negative skew distribution will be more symmetric if it is transformed as $r \rightarrow r^2$ with is correlation value [15].

Skewness of correlation value distributions involving GM and LCFM before and after transformed as follows.

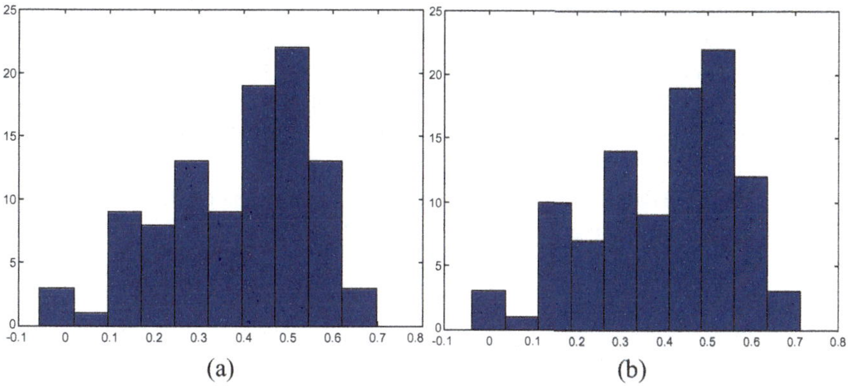

Fig. 3 Histograms of correlations: **a** GM and **b** LCFM

Table 2 Skewness of correlation value distributions

Correlation values involving:	Skewness	
	Before transformed	After transformed
GM	−0.613	0.259
LCFM	−0.577	0.255

Table 3 Confidence intervals for μ_{r^2}

Correlation values using:	Confidence interval (95%)
GM	$0.156 < \mu_{r^2} < 0.203$
LCFM	$0.162 < \mu_{r^2} < 0.209$

There could be seen from Table 2 that correlation values involving GM and LCFM become more symmetric (since they were close to zero).

Furthermore, we determined confidence interval for μ_{r^2} which is the average of the square of the correlation value of the model, each of which use GM and LCFM. The confidence intervals can be seen in the table below.

From Table 3, the average value of the squared correlation of the model using LCFM is in the interval (0.162, 0.209). This means that the correlation is still not good because we expect correlation value is more than 0.6. However, when compared with the model using GM, the average value of the correlation of the model using LCFM is even better.

To compare the results of the estimated corn weight obtained from each model, we took one test population. The correlation value of the model using GM to this test population is 0.697 (the maximum correlation value of the model using GM). For models using GM and LCFM, the results are presented in Table 4.

From Table 4, y^* is true corn weight, $\hat{y}^*_{GM}$ is estimated corn weight obtained from model using GM, $\hat{y}^*_{LCFM}$ is estimated corn weight obtained from model using LCFM, $y^* - \hat{y}^*_{GM}$ is the difference between true corn weight and estimated corn weight obtained from model using GM, while $y^* - \hat{y}^*_{LCFM}$ is the difference between true corn weight and estimated corn weight obtained from model using LCFM.

We obtained residual sum of squares of the estimated corn weight from model using GM is 20.893 while for model using LCFM is 20.137. So the estimated corn weight from model using GM is almost the same as the model using LCFM. For more details, there can be made also in the plot between y^* with each $\hat{y}^*_{GM}$ and $\hat{y}^*_{LCFM}$ as follows.

From Fig. 4, we can see that the estimates generated by the model using GM and the model using LCFM were almost the same. This was due to the GM which is actually a LCFM where

$$\sigma_v^2 = \frac{1}{2\sum_{j=1}^{100} p_j(1 - p_j)} \tag{13}$$

Table 4 Estimated corn weight obtained from model using GM and LCFM

y^*	$\hat{y}^*_{GM}$	$\hat{y}^*_{LCFM}$	$y^* - \hat{y}^*_{GM}$	$y^* - \hat{y}^*_{LCFM}$	y^*	$\hat{y}^*_{GM}$	$\hat{y}^*_{LCFM}$	$y^* - \hat{y}^*_{GM}$	$y^* - \hat{y}^*_{LCFM}$
2.015	2.450	2.444	−0.435	−0.429	3.719	4.621	4.516	−0.902	−0.797
3.779	3.375	3.360	0.404	0.419	0.582	2.048	2.036	−1.466	−1.454
4.011	3.873	3.837	0.138	0.174	0.527	0.997	1.077	−0.470	−0.550
2.542	2.630	2.615	−0.088	−0.073	3.205	2.898	2.936	0.307	0.269
3.568	3.834	3.740	−0.266	−0.172	1.996	2.275	2.351	−0.279	−0.355
3.116	2.727	2.744	0.389	0.372	1.619	2.167	2.176	−0.548	−0.557
2.099	1.782	1.778	0.317	0.321	4.788	4.040	3.986	0.748	0.802
1.826	3.167	3.138	−1.341	−1.312	2.641	2.335	2.277	0.306	0.364
1.612	2.295	2.318	−0.683	−0.706	3.583	3.427	3.398	0.156	0.185
1.543	1.731	1.732	−0.188	−0.189	2.966	2.480	2.526	0.486	0.440
0.012	1.609	1.609	−1.597	−1.597	3.140	1.693	1.798	1.447	1.342
2.965	2.975	2.953	−0.010	0.012	0.790	2.217	2.205	−1.427	−1.415
1.896	1.369	1.395	0.527	0.501	1.722	3.028	3.007	−1.306	−1.285
4.184	2.558	2.598	1.626	1.586	3.636	2.217	2.265	1.419	1.371

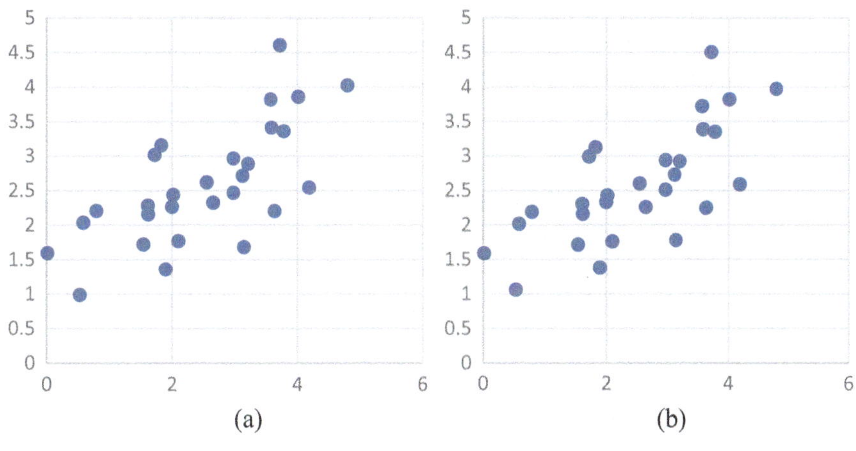

Fig. 4 Plot between true corn weight (horizontal) and estimated corn weight (vertical) for: **a** GM and **b** LCFM

Correlation between y^* and $\hat{y}^*_{GM}$ is 0.697 and correlation between y^* and $\hat{y}^*_{LCFM}$ is 0.711.

By using run test, the residuals $(y^* - \hat{y}^*)$ from both models (GM and LCFM) were tested for their randomness.

Hypothesis testing

Table 5 p-values for each residual from model using GM and LCFM

Model	The p-value
GM	0.359
LCFM	0.325

Table 6 Best correlations for model using GM and LCFM

Model with the best correlation	Correlation values	
	GM	LCFM
GM	0.562	0.553
LCFM	0.314	0.419

H_0: The residuals are random
H_1: The residuals are not random.

The p-values for each residual of the two models can be seen in Table 5.

Of the two models, we obtained p-values more than 30%. For a significance level less than 30%, H_0 is not rejected. The residuals of the two models can be said to be random. The two models were good enough.

Furthermore, the number of individuals used as a training population was reduced from 500 individuals to 50 individuals. Number of individuals in the test population was also reduced from 28 individuals to 5 individuals. A total of 100 runs were done. At each run, the correlation between y^* and $\hat{y}^*$ was calculated for the two models simultaneously. From these 100 runs, the best correlations for each model were presented in Table 6.

From Table 6, we see that the best correlation values for two models were 0.562 (GM) and 0.419 (LCFM). The best correlation between the two models was achieved by model using GM which was 0.941. At the time of the correlation value for the model using GM reached the maximum, the correlation value for the model using LCFM was 0.553. It was quite similar, however, when the correlation value for the model using LCFM is 0.419, the correlation value for the model using GM was 0.314. This indicated that there was not quite difference between model using GM and model using LCFM.

After that, we conducted the simulation to generate y (corn weight) using two models. Corn weight of 528 individuals was generated through the model $y = X\beta + \delta$ with $\delta \sim nMVN(\mathbf{0}, K' = K + \sigma_\varepsilon^2 I)$. Matrix X was taken from the data, β was estimated by b through least square method, and δ was generated from multivariate normal distribution. K was the covariance function matrix for each model (GM and LCFM). Matrix K was constructed from the data while σ_ε^2 was estimated numerically. Here are histograms of y for each model (Fig. 5).

It is seen that the distribution of y generated by the model using GM has two peaks. This is suitable with the data that consists of two populations. They are populations of corn weight grown in wet and dry land. It shows that this model is more suitable for the data. This is contrast with distribution of y generated by the model using LCFM which only has one peak. It does not correspond to the actual data.

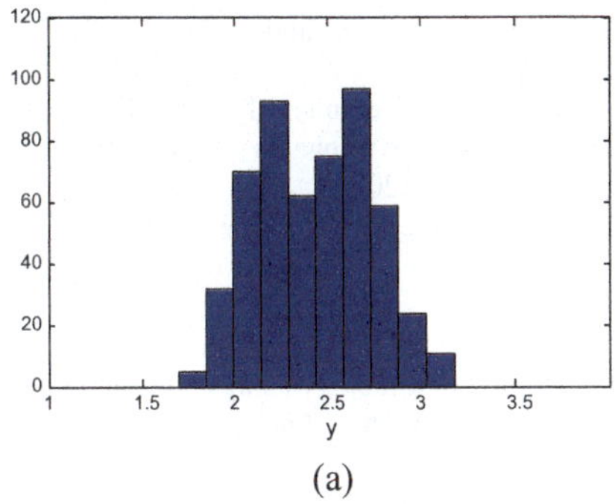

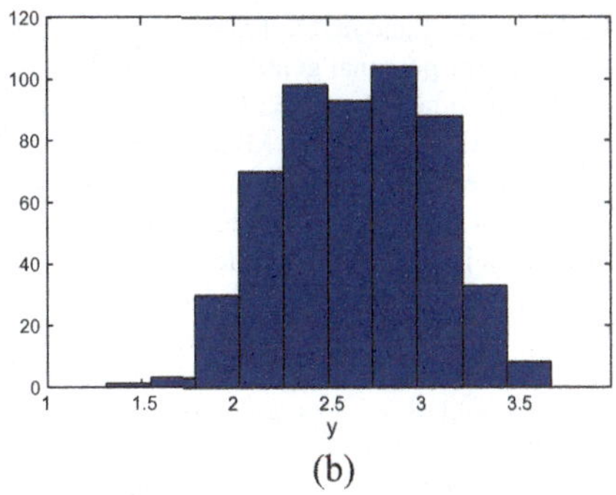

Fig. 5 Histograms of y for: **a** GM and **b** LCFM

5 Conclusion

In this paper, the problems assessed are: (1) the estimation of corn weight using mixed model which involved genomic relationship matrix (GM) and linear covariance function matrix (LCFM) and (2) the correlation between estimated corn weight and true corn weight for each model. A total of 500 genotyped individual corn (*Zea mays*) data were taken randomly from the population which consisted of 528 genotyped individual corn data. These data were used as training data while the rest which consisted of 28 genotyped individual corn data were used as test data. The correlation between estimated corn weight and true corn weight was computed for each model. The results obtained are: (1) The corn weight can be estimated using mixed model which involved genomic relationship matrix (GM) and linear covariance function matrix (LCFM) and (2) the 95% confidence interval for squared correlation between estimated corn weight and true corn weight from model using LCFM was 0.162–0.209 better than GM model which was 0.156–0.203.

References

1. Legarra, A., Christensen, O.F., Aguilar, I., Misztal, I.: Single step, a general approach for genomic selection. Livestock Sci. **166**, 54–65 (2014)
2. Ziegler, A., Bohossian, N., Diego, V.P., Yao, C.: Genetic prediction in the genetic analysis workshop 18 sequencing data. Genet. Epidemiol. **38**, S57–S62 (2014)
3. te Pas, M.F.W., Madsen, O., Calus, M.P.L., Smits, M.A.: The importance of endophenotypes to evaluate the relationship between genotype and external phenotype. Int. J. Mol. Sci. **18**, 472–493 (2017)
4. Yu, J., Dossa, K., Wang, L., Zhang, Y., Wei, X., Liao, B., Zhang, X.: PMDBase: a database for studying microsatellite DNA and marker development in plants. Nucleic Acids Res. **45**, D1046–D1053 (2017)
5. Nielsen, R., Paul, J.S., Albrechtsen, A., Song, Y.S.: Genotype and SNP calling from next-generation sequencing data. Nat. Rev. Genet. **12**, 443–451 (2011)

6. Karki, R., Pandya, D., Elston, R.C., Ferlini, C.: Defining "mutation" and "polymorphism" in the era of personal genomics. BMC Med. Genomics **8**, 1–7 (2015)
7. de Almeida Filho, J. E., Guimaraes, J. F. R., e Silva, F. F., de Resende, M. D. V., Munoz, P., Kirst, M., Resende Jr, M. F. R.: The contribution of dominance to phenotype prediction in a pine breeding and simulated population. Heredity **117**, 33–41 (2016)
8. Sargent, D.J., Geibel, M., Hawkins, J.A., Wilkinson, M.J., Battey, N.H., Simpson, D.W.: Quantitative and qualitative differences in morphological traits revealed between diploid *Fragaria* species. Ann. Bot. **94**, 787–796 (2004)
9. Zhou, H., Wang, X., Huang, Z., Gao, J., Guo, Y., Du, Y., Hu, H.: Identification of quantitative trait loci for fruit weight, soluble solids content, and plant morphology using an introgression line population of *Solanum pennellii* in a fresh market tomato inbred line. Horticultural Plant J. **2**, 26–34 (2016)
10. Su, G., Christensen, O.F., Janss, L., Lund, M.S.: Comparison of genomic predictions using genomic relationship matrices built with different weighting factors to account for locus-specific variances. J. Dairy Sci. **97**, 6547–6559 (2014)
11. Taylor, J.F.: Implementation and accuracy of genomic selection. Aquaculture **420–421**, S8–S14 (2014)
12. Knight, M.I., Daetwyler, H.D., Hayes, B.J., Hayden, M.J., Ball, A.J., Pethick, D.W., McDonagh, M.B.: An independent validation association study of carcass quality, shear force, intramuscular fat percentage and Omega-3 polyunsaturated fatty acid content with gene markers in Australian lamb. Meat Sci. **96**, 1025–1033 (2014)
13. Vanraden, P.M.: Efficient methods to compute genomic predictions. J. Dairy Sci. **91**, 4414–4423 (2008)
14. Crossa, J., de los Campos, G., Perez, P., Gianola, D., Burgueno, J., Araus, J. L., Makumbi, D., Singh, R. P., Dreisigacker, S., Yan, J., Arief, V., Banziger, M., Braun, Hans-Joachim: Prediction of genetic values of quantitative traits in plant breeding using pedigree and molecular markers. Genetics **186**, 713–724 (2010)
15. Feng, C., Wang, H., Lu, N., Chen, T., He, H., Lu, Y., Tu, X.M.: Log-transformation and its implications for data analysis. Shanghai Archives of Psychiatry **26**, 105–109 (2014)

Design and Implementation of Automatic Weather Station Using MQTT Protocol

Ida Wahyuni, Faddli L. Wibowo, Fandisya Rahman, Wayan Firdaus Mahmudy, and Atiek Iriany

Abstract Indonesia is a country with a large territory with various types of weather. However, the weather station in Indonesia is still not equipped with remote data transmission method automatically. Automatic weather station (AWS) which was developed in this study, has the ability to transmit weather data from measurement automatically using technology Message Queuing Telemetry Transport (MQTT) protocol. Based on the testing process by using the MQTT protocol on AWS, the packet data required in the data transmission is quite low, therefore the bandwidth requirement becomes lower as well. The bandwidth usage can be minimized by up to 64%. Moreover, the delay time required for data transmission using MQTT only 0.003 seconds. It is shorter when compared with HTTP which needs 16,658 seconds of delay time. Because of that, the data transmission becomes faster. It is proved that using MQTT on AWS can improve the process of sending weather data more efficiently.

Keywords Automatic weather station · HTTP protocol · MQTT protocol · Weather

1 Introduction

Measurement of weather data becomes a very important thing in Indonesia. That is because many sectors need weather data for a variety of needs, including the transportation sector, the agricultural sector [1, 2], and the energy sector. In the transportation sector, especially air and maritime transportation need rainfall, wind speed, and wind direction data. In the agricultural sector, weather data such as rainfall data are needed to determine the growing season, especially for root crops such as

I. Wahyuni (✉) · F. L. Wibowo · F. Rahman · W. F. Mahmudy
Department of Computer Science, Brawijaya University, Malang, Indonesia
e-mail: ida.wahyuni8@gmail.com

A. Iriany
Department of Mathematics and Natural Sciences, Brawijaya University, Malang, Indonesia

E. Joelianto et al. (eds.), *Cyber Physical, Computer and Automation System*, Advances in Intelligent Systems and Computing 1291,
https://doi.org/10.1007/978-981-33-4062-6_16

potatoes [3]. In other cases, weather data also need in the energy sector, because the network performance of electricity can be affected by several factors, including temperature, humidity, and weather [4].

Indonesia currently has a lot of weather stations, but its location is limited to certain areas. Many areas in Indonesia, especially in remote areas that do not have a weather gauge. Due to the wide area of Indonesia, it becomes a challenge to make the weather gauges that can be placed in a remote location and can send data automatically.

Automatic weather station (AWS) was once patented by Diamond and Hinman [5]. This AWS has many measurement data they are barometric pressure, ambient temperature and relative humidity, wind velocity and direction, rainfall, visibility, ceiling height, and other factors probably. The architecture of AWS by Diamond and Hinman [5] was used a keyed radio transmitter and an aural radio receiver to provide the transmission of meteorological data from a remote point.

There were a lot of researches that discussed the tools which can be used as measuring weather data, but at some earlier research data delivery mechanisms on the weather gauges still can transmit data over long distances [6–8]. A remote data delivery mechanism is needed to gauge the effectiveness of using the weather. With the method of delivery of weather data remotely, then the delivery of weather data to users becoming more actual and easy.

It required weather measuring instrument which has advantages in measurement and can make long-distance data transmission automatically. One instrument that can be made is automated weather gauges or automatic weather stations. Research on the weather gauges with weather data delivery system automatically uses the GSM network services through short message service (SMS) was made by Satria and Siregar [9]. This study describes the manufacture of rainfall gauges with a microcontroller using a GSM network. Instruments in the study were designed to send the rainfall data every 10 min with the smallest scale of 0.2 mm [9]. This study offers the design and manufacture of automated weather gauges (automatic weather station) with technology that provides more sophisticated and modern that is using the protocol Message Queuing Telemetry Transport (MQTT) for data transmission communication tool. Delivery of weather data using the General Packet Radio Service (GPRS).

However, the communication protocol used is protocol MQTT that has the capability of sending data better and is supported by the quality of service (QoS) is the ability of a network to provide better service again for traffic through it so that the data sent is certainly up [10]. More MQTT protocol also a small packet of his data, so less bandwidth is used. In addition, the MQTT protocol can also be used with a resource or a small power [11]. So that the communication protocol is better for data transmission, so the weather data transmission can also be more quickly and effectively even to remote areas in Indonesia strip.

This research will develop an automatic weather station (AWS) that can collect five weather data they are rainfall, wind direction, wind speed, temperature, and humidity. This AWS will develop with MQTT protocol. This research also does the performance testing between HTPP and MQTT protocol and they are bandwidth testing, delay testing, and packet size testing to see the different performance of this protocol.

2 Weather Station Component

The components that will be used in the development process of an automatic weather station (AWS) include a weather station tool, microcontroller, Mini SIM900A, and protocol MQTT for communication.

2.1 Weather Station

The weather station is used as a measurement tool to obtain weather data. The weather has six elements they are temperature, relative humidity, precipitation or rainfall, wind speed, wind direction, and atmospheric pressure [7]. The weather measuring instruments all on the weather station package. The measuring instruments of the weather station are wind vane module, anemometer module, rain gauge, pressure sensors (BMP 180), sensor temperature and humidity (am 2320), and weather link module (bht4dv). The wind vane module is used to determine the wind direction. Module anemometer is used to determine wind speed. Module rain gauge is used to measure water precipitation or rainfall [12]. The pressure sensor is used to measure air pressure. DHT 11 temperature sensors are used to measure temperature and humidity. Pictures of each component weather station can be seen in Figs. 1, 2, 3, and 4.

2.2 Microcontroller

The microcontroller is an electronic device built from an integrated circuit (IC) that has the ability to manipulate data or information based on a sequence of program instructions created [9]. The structure of the microcontroller is equipped with supporting components such as a processor, flash memory, clock, and so forth. One microcontroller that is often used is Arduino with a microcontroller development board that uses an automatic voltage regulator (AVR) [6].

2.3 Protocol Message Queuing Telemetry Transport (MQTT)

As already mentioned, protocol Message Queuing Telemetry Transport (MQTT) stands for MQ Telemetry Transport [13]. MQTT protocol is an application layer protocol designed for constrained devices, lightweight messaging protocol, and very

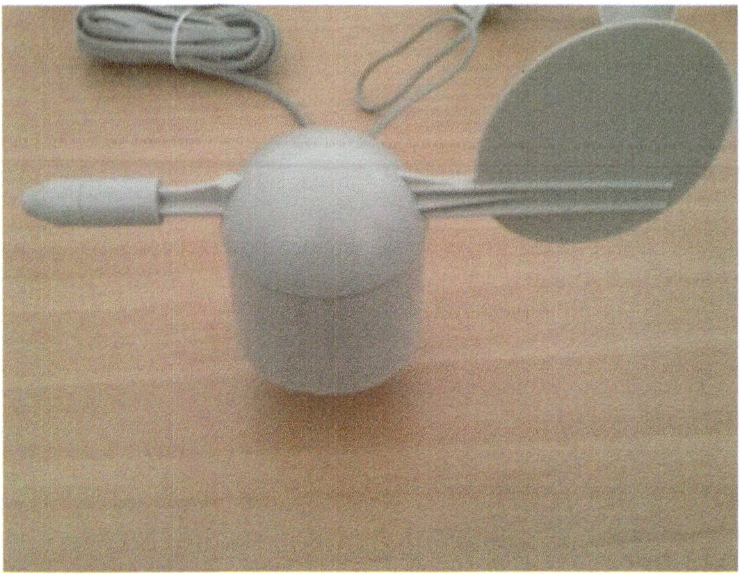

Fig. 1 Wind vane module

Fig. 2 Anemometer module

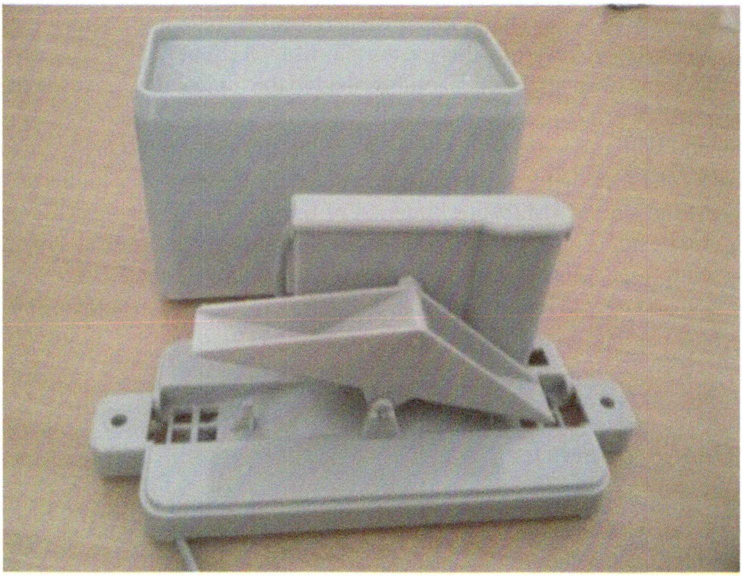

Fig. 3 Rain gauge module

Fig. 4 Weather link module with pressure sensor, temperature, and humidity sensor

simple, high latency or unreliable networks and low bandwidth. It uses a topic-based publish-subscribe architecture, this means that when a client publishes a message to a particular topic, then all the clients subscribed to the topic will receive the message [14]. The design principles are to minimize network bandwidth and device resource requirements while also attempting to ensure reliability and some degree of assurance of delivery. These principles also turn out to make the protocol ideal of the emerging "machine-to-machine" (M2M) or "Internet of things" world of connected devices, and for mobile applications where bandwidth and battery power are at a premium.

This study used MQTT because the protocol has the characteristics of simple, lightweight, and easy to implement. Another reason is MQTT can be used on devices with limited computing resources, such as Arduino [10].

2.4 Quality of Service (QoS)

The quality of service (QoS) is the ability of a network to provide better service for traffic that passes through so that the data sent is certainly up [10]. QoS is composed of three parts, namely delay, jitter, and bandwidth. The delay is a time delay that is owned by a package that is caused by the transmission from one point to another destination. Jitter is a kind of delay that occurs on the network Internet Protocol (IP). While bandwidth is a wide frequency used by the signal to transmit data packets.

3 Implementation

3.1 Design of Automatic Weather Station

Automatic weather station (AWS) is made with several supporting components include Arduino, GMS SIM900A module, MQTT server, Node JS, and databases in this case MongoDB. Those components are assembled into a single unit that produces tools for weather station capable of transmitting weather data automatically. Architecture automatic weather station (AWS) is shown in Fig. 5. The automatic weather station (AWS) that has been designed has three meters high and the detail is shown in Fig. 6.

Publisher The publisher is the node that sends information to the broker's server on a regular basis that the information will be read by subscribers [11]. The publisher consists of several components, namely *weather link* weather station is used as a measure of air pressure, temperature, and humidity, *rain gauge* module that is used to measure rainfall, *wind direction* module which is used as a direction of the wind, and the *wind speed* module is used as a measure of the wind speed.

Subscriber The subscriber is a client that receives data from publishers, subscribers will never get a publishing message if the client does not subscribe to any

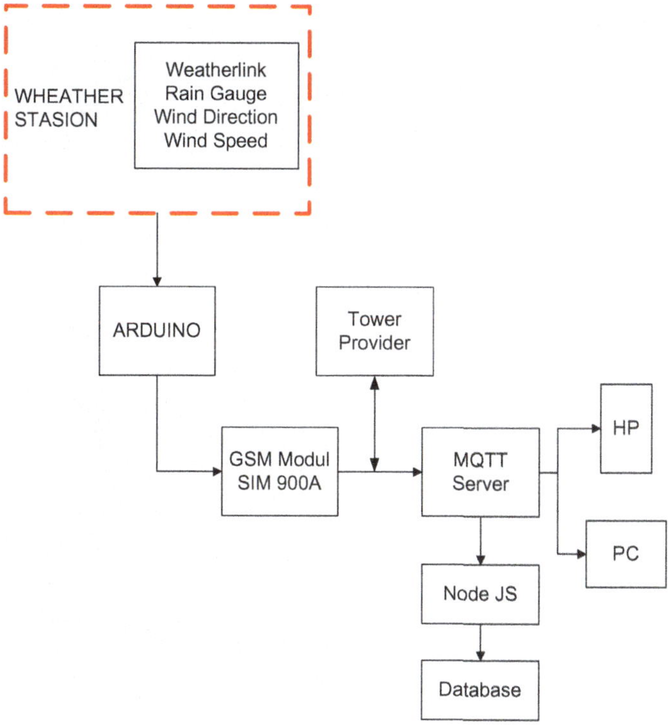

Fig. 5 Architecture of automatic weather station

topic [11]. A client needs to send a subscribe message to the MQTT broker in order to receive relevant messages. The communication model rule of the publisher or subscriber is that components which are interested in consuming certain information register their interest.

Arduino Arduino is a microcontroller development board that uses an automatic voltage regulator (AVR) [6]. This study used Arduino to control or read data from *weather link* then instructs the SIM900A GSM module to transmit data to the server MQTT. The purpose of ease of use and simplicity in programming because downloader modules into a single board with a microcontroller so users do not need to make the circuit more to fill the program.

GSM SIM900A Module GSM/GPRS connection works only in the presence of reliable network coverage, a condition often difficult to reach in the high mountain [15]. GSM module has the following features such as support dual frequency of GSM900 and GSM1800, compact and low power consumption, the reliable and safe transmission of data, provide fast, provide standard AT command interface to users, voice, short message, and fax [16]. It is ideal for this system because of its high quality short message function. In this study, the GSM SIM 900A module serves as a means for sending data from the weather station to the MQTT server with GPRS mobile networks. Before the data is received by the MQTT server, the GSM module

Fig. 6 Automatic weather station

requires an assistance provider for the data transmission process. This research is not testing about the ability of the GSM module if it located in a high mountain or remote place.

MQTT Server As mention before, Message Queue Telemetry Transport (MQTT) is an open protocol designed by IBM, was originally intended for unreliable networks with restricted resources such as low bandwidth and high latency [17]. The MQTT server to function as a protocol as described in the literature review. MQTT server will coordinate and collect weather data obtained from the weather station. MQTT is used for communication between publisher and subscriber [18].

Node JS Node JS is a JavaScript framework that is designed to be suitable for the environment are IoT and wireless sensor network which has the properties focusing on performance and low memory consumption [19]. In this study, Node JS is used as a bridge or a bridge that connects the data from the MQTT server to the database.

Database Databases appear as a repository with organized and structured data, where all that the data is combined into a set of registers arranged into a regular structure to enable easy extraction of information, one type of database that is commonly

Fig. 7 Li–ion polymer U30GT battery

used today is Mongo DB [20], and this study also used a database Mongo DB. Mongo DB database used as a storage area weather data obtained from the weather station.

3.2 Power Supply for Automatic Weather Station

The power supply used to run the automatic weather station (AWS) is a U30GT lithium-ion polymer battery, with a capacity of 12,000 mAh, 7.4 V output voltage, input voltage of 12 V, 3.8 × 151 × 125 mm size, and solar cell 18 V 10 W. Battery and solar cell employed is shown in Figs. 7 and 8.

4 Testing Phase and Result

4.1 Testing Accuracy of Automatic Weather Station

The testing process is performed to determine the weather data in the city of Malang. Locations testing performed outdoors or in the field, exactly done in the area of Soekarno Hatta Street, Malang, East Java, by putting AWS at an altitude of 10 m above the ground. Measurement data is verified with the results of weather measurements from the official Web site of the Meteorology, Climatology, and Geophysics (BMKG) Indonesia is www.bmkg.go.id. The testing process conducted over two days from the time range at 04.00–12.00 PM, with a delivery time of data every 5 min.

Fig. 8 Solar cell for AWS

The test is conducted in order to test the ability of the instrument to measure the weather, other than that this test will also feature the ability to send data from the location of the placement of automatic weather station (AWS) to a server that is a mobile phone or personal computer (PC). The result of testing the AWS is shown in Tables 1 and 2. While the weather data obtained from BMKG predicted results are shown in Table 3.

The testing results in Tables 1 and 2 show that all the components of the automatic weather station (AWS) are function properly. The average of rainfall, temperature, humidity, and air pressure is also on a range of measurement data from BMKG. This indicates that the AWS performs with the right measurements of weather data.

4.2 Bandwidth Testing

Bandwidth testing is done to determine how much bandwidth is required for sending data from AWS to the server. Bandwidth testing is done with Wireshark software in a way to capture the data bandwidth within a certain timeframe. The time used to capture is every 10 min, this was done because the implementation time data is sent to the server for 10 min/send. The bandwidth testing is tested on a different day but it used the same time from 6:10 – 7:40 AM. The results of bandwidth testing are shown in Table 4.

Table 1 Weather measurement results with AWS at September 10, 2016

Id	Date	Time	*WD*	WS_{min}	WS_{max}	RH	RD	T	H	P
1	9/10/16	16:46:26	315	0	2.24	4.83	4.83	31.67	57	2.3
2	9/10/16	16:51:41	90	0	1.34	4.83	4.83	31.11	58	2.3
3	9/10/16	16:56:51	0	0	1.79	0.25	4.83	30.56	59	2.3
4	9/10/16	17:02:01	45	0	1.79	0	4.83	30.56	60	2.3
…	…	…	…	…	…	…	…	…	…	…
65	9/10/16	23:36:43	315	0	0.45	0	4.83	22.78	74	2.3
66	9/10/16	23:41:55	270	0	0.45	0	4.83	22.78	74	2.3
67	9/10/16	23:47:05	180	0	0	0	4.83	22.78	73	2.3
68	9/10/16	23:52:16	225	0	0.89	0	4.83	22.78	74	2.3
69	9/10/16	23:57:29	225	0	0.89	0	4.83	22.78	74	2.3

Note Wind direction (*WD*–degree), Wind speed minimum (WS_{min}–m/s), Wind speed maximum (WS_{max}– m/s), Rain one hour (RH - mm/h), Rain one day (RD-mm/day), Temperature (T–°C), Humidity (H–%), Pressure (P–bar)

Table 2 Weather measurement results with AWS at September 11, 2016

Id	Date	Time	*WD*	WS_{min}	WS_{max}	RH	RD	T	H	P
1	9/11/16	16:49:51	225	0	0	0	0	26.67	64	2.3
2	9/11/16	18:12:07	270	0	0	0	0	27.22	67	2.3
3	9/11/16	18:17:16	315	0	1.34	2.29	2.29	30	63	2.3
4	9/11/16	18:22:30	270	0	1.34	2.29	2.29	29.44	63	2.3
…	…	…	…	…	…	…	…	…	…	…
44	9/11/16	1:10:28	315	0	0	0	4.83	23.89	74	2.3
45	9/11/16	1:15:22	315	0	0	0	4.83	23.89	74	2.3
46	9/11/16	1:20:36	315	0	0	0	4.83	23.89	74	2.3
47	9/11/16	1:25:47	315	0	0.45	0	4.83	23.89	74	2.3
48	9/11/16	1:30:57	315	0	0.45	0	4.83	23.89	74	2.3

Table 3 Weather data from BMKG prediction results

Date	Rain	Wind speed max (km/h)	Temp (C)	Hum
9/10/2016	0	0.5–3.0	20 - 30	45 - 88
9/11/2016	0	0.5–3.5	19 - 31	45 - 88

From Table 4, it can be seen that the uses of the bandwidth of the MQTT protocol are much smaller when compared to the HTTP protocol. According to the test results, the average bandwidth used in the MQTT protocol is 779.8 that number is quite small when compared to the average bandwidth used by the HTTP protocol, which

Table 4 Bandwidth testing results (bytes)

MQTT protocol		HTTP protocol	
Time	Bandwidth usage	Time	Bandwidth usage
6:10:00 AM	790	6:10:00 AM	2152
6:20:00 AM	844	6:20:00 AM	2148
6:30:00 AM	730	6:30:00 AM	2148
6:40:00 AM	844	6:40:00 AM	2152
6:50:00 AM	844	6:50:00 AM	2156
7:00:00 AM	724	7:00:00 AM	2088
7:10:00 AM	724	7:10:00 AM	2088
7:20:00 AM	610	7:20:00 AM	2148
7:30:00 AM	844	7:30:00 AM	2148
7:40:00 AM	844	7:40:00 AM	2148
Average	779.8	Average	2137.6

reached 2137.6. It shows that by using the MQTT protocol, the bandwidth usage can be minimized up to 64%.

4.3 *Delay Testing*

Delay tests are performed to determine how high the lag time occurred to send data from AWS to the server. Delay tests obtained using the same software for testing bandwidth which namely as Wireshark software, it can capture the delay time within a certain timeframe. Time to capture is set every 10 min because the data is sent to the server every 10 min. The way of delay tests is performed can be described as follows, first, the client time will be synchronized with the server time, every time the client sends a packet then it recorded the current time (e.g., 10:20:21), and then it recorded the time when the package arrived on the server side (e.g., 10:20:50). The above examples indicate that the delay time is 29 s. The result of the testing delay using MQTT and HTTP is shown in Table 5.

The results of delay testing in Table 5 indicate that the delay with the MQTT protocol much smaller when compared to the HTTP protocol. The average time delay using MQTT only 0.00346 s, it is much smaller when compared to the time delay using HTTP which the average is reaching 16.6587 s. From these tests, it can be concluded that the use of the MQTT protocol more effective and efficient in the delivery process of the data.

Table 5 Delay testing results (seconds)

MQTT protocol		HTTP protocol	
Time	Delay	Time	Delay
6:10:00 AM	0.0035	6:10:00 AM	12.39
6:20:00 AM	0.0035	6:20:00 AM	19.21
6:30:00 AM	0.0035	6:30:00 AM	5.24
6:40:00 AM	0.0027	6:40:00 AM	26.66
6:50:00 AM	0.0028	6:50:00 AM	5.097
7:00:00 AM	0.0031	7:00:00 AM	20.57
7:10:00 AM	0.0035	7:10:00 AM	23.06
7:20:00 AM	0.0027	7:20:00 AM	22.10
7:30:00 AM	0.0065	7:30:00 AM	14.00
7:40:00 AM	0.0028	7:40:00 AM	18.26
Average	0.00346	Average	16.6587

4.4 *Packet Size Testing*

Packet size testing is performed to determine the size of the required packages in the shipping process of the data on the AWS device. The test uses Wireshark software, which can capture the data packet size within a certain time. It captures every 10 min, this is because the data need to be sent to the server every 10 min. The test results of packet size are shown in Table 6.

The test results indicated that the packet size required by using MQTT is quite small. The average use of packet size using MQTT is 11.2, it is much smaller when compared when using HTTP which the average reaching 33.8. The size of the packet

Table 6 Packet size testing results

MQTT protocol		HTTP protocol	
Time	Delay	Time	Delay
6:10:00 AM	11	6:20:00 AM	34
6:20:00 AM	12	6:20:00 AM	34
6:30:00 AM	10	6:30:00 AM	34
6:40:00 AM	12	6:40:00 AM	34
6:50:00 AM	12	6:50:00 AM	34
7:00:00 AM	11	7:00:00 AM	33
7:10:00 AM	11	7:10:00 AM	33
7:20:00 AM	9	7:20:00 AM	34
7:30:00 AM	12	7:30:00 AM	34
7:40:00 AM	12	7:40:00 AM	34
Average	11.2	Average	33.8

size affects the delay and bandwidth usage. The smaller the packet size required a smaller time delay and bandwidth [21].

5 Conclusion

Based on the testing results on AWS which was developed to take weather measurements, this study proves that by using MQTT protocol the ability to transmit weather measurement data is better than using HTTP. Based on the testing process by using MQTT protocol on AWS, the packet data required in the data transmission is quite small, therefore the bandwidth requirement becomes smaller as well. Moreover, the delay time required for data transmission is shorter than before, which makes the data transmission faster. It is proved that using MQTT on AWS can improve the process of sending weather data more efficiently. For next study, an AWS will be embedded with software for rainfall prediction using many methods that have been developed before such as Tsukamoto FIS [22], Hybrid Tsukamoto FIS and GA [1], or Hybrid ANFIS and GA [2].

References

1. Wahyuni, I., Mahmudy, W.F.: Rainfall prediction in tengger-indonesia using hybrid tsukamoto FIS and genetic algorithm. J. ICT Res. Appl. **11**(1), 38–54 (2017)
2. Wahyuni, I., Mahmudy, W.F., Iriany, A.: Rainfall prediction using hybrid adaptive neuro fuzzy inference system (ANFIS) and genetic algorithm. J. Telecommun. Electron. Comput. Eng. **9**(2–8), 51–56 (2017)
3. Rosyidah, A., Wardiyati, T., Abadi, A.L., Maghfoer, M.D.: Enhancement in effectiveness of antagonistic microbes by means of microbial combination to control ralstonia solanacearum on potato planted in middle latitude. AGRIVITA **35**(2), 174–183 (2013)
4. Suryani, E., Hendrawan, R.A., Adipraja, P.F.E., Dewi, L.P.: A simulation model for strategic planning in asset management of electricity distribution network. In: 4th International Conference Software Computing Intelligence Systems Information Technology, vol. 516, pp. 539–550. (2015)
5. Diamond, H., Hinman, W.S.: Automatic Weather Station (1942)
6. Wicaksono, A.S.: Design and implementation local wheather station powered by solar cells. Faculty Electrical Engineering Telkom University, pp. 1–53. (2015)
7. Olatomiwa, L.J., Adikwu, U.S.: Design and construction of a low cost digital weather station. AU J. Technol. **16**(2), 125–132 (2012)
8. Parvez, S.H. et al.: Design and implementation of a cost effective, portable and scalable electronic weather station. Recent Res. Electri. Comput. Eng. Weather May 2016 (2015)
9. Satria, B., Siregar, P.: Otomatisasi Penakar Hujan dengan Mikrokontroler Menggunakan Jaringan GSM (Automation of raincatcher with microcontroller using GSM network). Megasains **3**(2), 97–106 (2012)
10. Tarigan, S.O.F., Sitepu, H.I., Hutagalung, M.: Pengukuran Kinerja Sistem Publish/Subscribe Menggunakan Protokol MQTT (Message Queuing Telemetry Transport) (Publish / Subscribe System performance measurement using the MQTT (message queuing telemetry transport protocol). J. Telemat. **9**(1), 25–30 (1858)

11. Hunkeler, U., Truong, H.L., Stanford-Clark, A.: MQTT-S—A publish/subscribe protocol for wireless sensor networks. In: 2008 3rd International Conference Communications Systems Software Middle Work (COMSWARE '08), pp. 791–798. (2008)
12. Roussey, C., et al.: Weather station data publication at Irstea: an implementation report. CEUR Workshop Proc. **1401**, 89–104 (2014)
13. Message Queuing Telemetry Transport (MQTT) (2016). https://mqtt.org
14. Thangavel, D., Ma, X., Valera, A., Tan, H.X., Tan, C.K.Y.: Performance evaluation of MQTT and CoAP via a common middleware. In: IEEE ISSNIP 2014-2014 IEEE 9th International Conference Intelligence Sensors, Sensors Networks Information Processing Confernce Proceedings no. April, pp. 21–24. (2014)
15. Abbate, S., Avvenuti, M., Carturan, L., Cesarini, D.: Deploying a communicating automatic weather station on an alpine glacier. Procedia Comput. Sci. **19**, 1190–1195 (2013)
16. Huang, H., Xiao, S., Meng, X., Xiong, Y.: A remote home security system based on wireless sensor network and GSM technology. In: NSWCTC 2010-2nd International Conference Networks Security Wireless Communications Trust Computer, vol. 1, pp. 535–538. (2010)
17. Lee, S., Kim, H., Hong, D.K., Ju, H.: Correlation analysis of MQTT loss and delay according to QoS level. In: International Confernce Information Network, pp. 714–717. (2013)
18. De Caro, N., Colitti, W., Steenhaut, K., Mangino, G., Reali, G.: Comparison of two lightweight protocols for smartphone-based sensing. In: IEEE SCVT 2013-Proceedings 20th IEEE Symposium Communications Vehicle Technology BeNeLux, pp. 1–6. (2013)
19. Tilkov, S., Verivue, S.V.: Node . js : using javascript to build high-performance network programs. IEEE Internet Comput. **7801** (2010)
20. Abramova, V., Bernardino, J.: NoSQL databases: MongoDB vs cassandra. In: Proceedings International C* Conference Computer Science Software Engineering ACM 2013, pp. 14–22. (2013)
21. Dhar, P., Gupta, P.P., College, G.H.R.: Intelligent parking cloud services based on IoT using MQTT protocol. Int. J. Eng. Res. **5013**(5), 457–461 (2016)
22. Wahyuni, I., Mahmudy, W.F., Iryani, A. Rainfall prediction in tengger region indonesia using tsukamoto fuzzy inference system. In: 1st International Conference Information Technology Information Systems Electrical Engineerning ICITISEE 2016, vol. 16, pp. 130–135. (2016)

Rainfall Prediction in Tengger Indonesia: A System Dynamics Approach

Ida Wahyuni, Philip Faster Eka Adipraja, Wayan Firdaus Mahmudy, and Atiek Iriany

Abstract Commonly, rainfall prediction uses historical data or the time series data. But there are many factors can affect rainfall such as temperature and humidity. Therefore, an approach is needed that can model forecasting rainfall not only uses time series data but also includes other supporting variables. System dynamics is an approach for modeling and simulation of the data. By modeling the rainfall data from the previous year which includes temperature and humidity data using system dynamics, prediction of rainfall can be simulated. The prediction results using systems dynamics approach generate root mean square error (RMSE) smaller than previous studies. The result of RMSE with system dynamics for Puspo area is 6.767. It is smaller than the RMSE results with a hybrid Tsukamoto FIS with genetic algorithm method which is 7.30.

Keywords Indonesia · Prediction · Rainfall · System dynamics · Tengger

1 Introduction

Today's rainfall prediction becomes very important, that is because climate change is affecting rainfall patterns [1]. In recent years, the pattern of rainfall becomes more erratic and quite difficult to predict. Besides climate change, many other factors can influence rainfall rate. Some of them are the air temperature, air humidity, the air pressure difference, sunlight, and water vapor [2].

Research on rainfall prediction has been done by various methods. One simple method that has been conducted for rainfall forecasting is GSTAR-SUR [1] to predict rainfall in the Tengger, but the largest RMSE in the area Tutur is 10.89. Another study conducted in [3] using methods Tsukamoto FIS produces smaller RMSE which

I. Wahyuni (✉) · W. F. Mahmudy · A. Iriany
Department of Computer Science, Brawijaya University, Malang, Indonesia
e-mail: ida.wahyuni8@gmail.com

I. Wahyuni · P. F. E. Adipraja
Department of Informatics, Institut Teknologi dan Bisnis Asia, Malang, Indonesia

E. Joelianto et al. (eds.), *Cyber Physical, Computer and Automation System*, Advances in Intelligent Systems and Computing 1291,
https://doi.org/10.1007/978-981-33-4062-6_17

is 8.64 at Tutur area. Another method uses Hybrid Tsukamoto FIS, and genetic algorithm has also been conducted in [4] to predict rainfall in the same location with smaller RMSE which is 6.63

Many methods used for forecasting rainfall use time series data only [1, 3, 4], though many factors influence rainfall such as temperature and humidity of the area. If other factors that affect the rainfall are calculated, the prediction results most likely will be more accurate.

Therefore, a simple method is needed that can predict rainfall with lots of input variables. System dynamics is an approach that is able to analyze and design a policy that is characterized by dependence, mutual interaction, information feedback, and the causal loop [5]. System dynamics approach can be used as a forecasting simulation method that has been done in many previous studies. System dynamics is once used by Dyson and Chang [6] for forecasting municipal solid waste generation in a fast-growing urban region. In the study, Dyson and Chang [6] findings clearly indicate that such a new forecasting approach may cover a variety of possible causative models and track inevitable uncertainties down when traditional statistical least squares regression methods are unable to handle such issues.

This study uses a system dynamics approach to perform rainfall prediction using data in the past or the time series data and some of weather variable. Input variables used for this study is the rainfall data in Tengger, East Java, Indonesia from 2005 to 2014. Moreover, this approach uses other input variables such as temperature and humidity as the factors that influence rainfall.

2 Literature Review

2.1 Rainfall

Rainfall is the amount of rain that poured down in an area within a specified period. Rainfall data has the dynamic physical characteristics, where the data is taken within a specified time interval or time series. Rainfall data requires time series analysis because of large different of intensity between the dry season and the rainy season in a certain region. With time series analysis, intensity fluctuations can be analyzed and compared for every year [7].

The degree of rainfall is expressed by the amount of rainfall in a certain time unit, example mm/hour. In the world of meteorology, raindrops with diameters between 0.5–0.1 mm referred to drizzle and a diameter greater than 0.5 mm is called rain. Rain grain size is directly proportional to the speed of the fall of rain, the larger the size of raindrops, the greater the speed when it falls. The accuracy of rainfall measurement tool that exists today is 1/10 mm, when reading is done only once a day and recorded as rainfall previous day or the last day [8].

2.2 System Dynamics

System dynamics is an approach that can perform analysis and design of the policy with the help of computers. System dynamics is typically characterized by dependency, information feedback, interaction mutualism, and the causal loop [9].

System dynamics has been used since the 1960 s for modeling and simulating various problems. System dynamics approach begins with defining the problem and simulation goals, followed by process of mapping and modeling variables that influence in the model. If all the variables are considered sufficient and match with the goals, then develop stock and flow diagram (SFD) by giving the value or weight of each variable [10]. In this study, the use of system dynamics is only for modeling and simulation to predict rainfall.

The steps of simulation adapted from Sterman begins with setting goals and continued with the development of the model [11]. The simulation results of the developed model needs to be validated in two ways: mean comparison (E1) and variance comparison (E2) [12].

3 Simulation

3.1 Development of Model

Precipitation is also known as rainfall, snowfall, and other forms of liquid or frozen water that fall from the clouds. Precipitation occurs at irregular intervals, and the character of the rainfall clearly depends on the temperature and weather condition. When the air warms up above freezing point, the precipitation is turned into rain. The water holding capacity of air increases 6–7% for every increase in temperature of 1 °C [13].

According to the data from National Oceanic and Atmospheric Administration (NOAA), the temperature in Tengger shows a monthly pattern where the October is the warmest month throughout the year that the average temperature is around 28 °C. And the average temperature of July is about 26 °C which is the lowest average temperature throughout the year [14]. SFD diagram of monthly temperature is shown in Fig. 1.

From the temperature, data shown that each month has a different temperature ranges. Those data mergers into one variable which includes the temperature graph in one year. The value of a temperature variable calculated at each input uses two coefficient values which are the minimum and maximum temperature. Some equations used to calculate the variable temperature of each month are shown in Table 1.

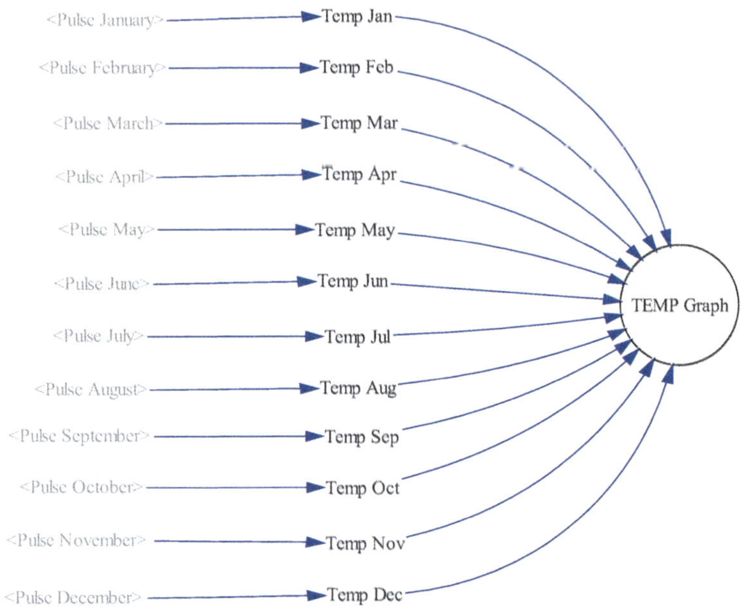

Fig. 1 SFD of temperature categorized by month

Table 1 Equation used to generate temperature variable each month

Variable	Equation
TEMP graph	= Temp Jan + Temp Feb + . . . + Temp Nov + Temp Dec Units: °C
Pulse January	= PULSE TRAIN (1, 1, 12, FINAL TIME) Units: dimensionless
. . .	. . .
Pulse December	= PULSE TRAIN (12, 1, 12, FINAL TIME) Units: dimensionless
Temp Jan	= IF THEN ELSE (Pulse January = 1, RANDOM UNIFORM(22.5, 30.3, 0), 0) Units: °C
. . .	. . .
Temp Dec	= IF THEN ELSE (Pulse December = 1, RANDOM UNIFORM (22.6, 31.2, 0), 0) Units: °C

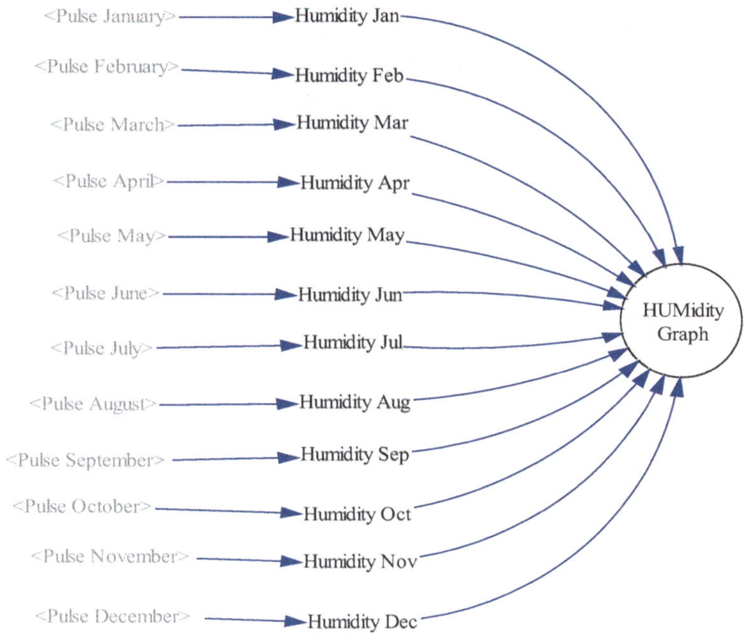

Fig. 2 SFD of humidity categorized by month

When air rises to the low pressure area, the air will expand and become cooler, and the cooling causes the water vapor to form dew and create rainfall. In this case, the temperature changes greatly affect the type and amount of precipitation through the humidity in the air. The humidity in the ground will increase 4.3% per 1 °C changes in temperature [4]. SFD of humidity in monthly categories can be seen in Fig. 2.

Each month has different humidity ranges, merger into a single variable that includes a graph of humidity in one year. The variable of humidity calculated on each input uses two coefficient which are the minimum and maximum humidity point. Some equations used to calculate the humidity variable of each month are shown in Table 2. However, the equation for pulses variable could be seen in Table 1.

In this study, using 10 years' historical rainfall data (January 2005–December 2014) was taken from four locations: Puspo, Sumber, Tosari, and Tutur, where the average rainfall in every month shows a similar pattern in every year. The monthly rainfall pattern can be seen in Fig. 3. Those locations where the rainfall data was taken rainfall show similar pattern.

Based on rainfall data in the previous years that have been acquired, the SFD of rainfall for each location needs to be categorized according to the month to obtain a monthly pattern simulation similar to the original pattern. SFD of rainfall from Puspo location can be seen in Fig. 4.

Each month has a different range of rainfall, which is merger into the variable that includes the graph of rainfall in a year. The RAW data obtained from Tengger area

Table 2 Equation used to generate humidity variable each month

Variable	Equation
Humidity graph	= Humidity Jan + Humidity Feb + ... + Humidity Nov + Humidity Dec Units: percent
Humidity Jan	= IF THEN ELSE(Pulse January = 1, RANDOM UNIFORM(97, 57, 0), 0) Units: percent
...	...
Humidity Dec	= IF THEN ELSE (Pulse December = 1, RANDOM UNIFORM (99, 45, 0), 0)) Units: percent

Fig. 3 Annual rainfall pattern

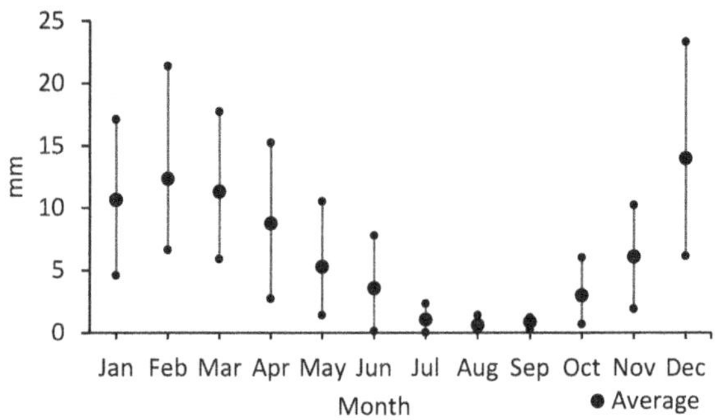

(Pospo, Sumber, Tosari, and Tutur) is categorized by month. Then retrieve average, maximum, minimum, and standard deviation from each month of the data. The value of rainfall calculated on each input uses four coefficient that is the minimum, maximum, standard deviation, and average of each month according to the data. Some equations used to calculate the monthly rainfall are shown in Table 3. However, the equation for pulses variable could be seen in Table 1.

Rainfall prediction using system dynamics uses two affecting variables such as temperature and humidity. SFD of the rainfall prediction can be seen in Fig. 5. Furthermore, some equations used to calculate the rainfall prediction are shown in Table 4.

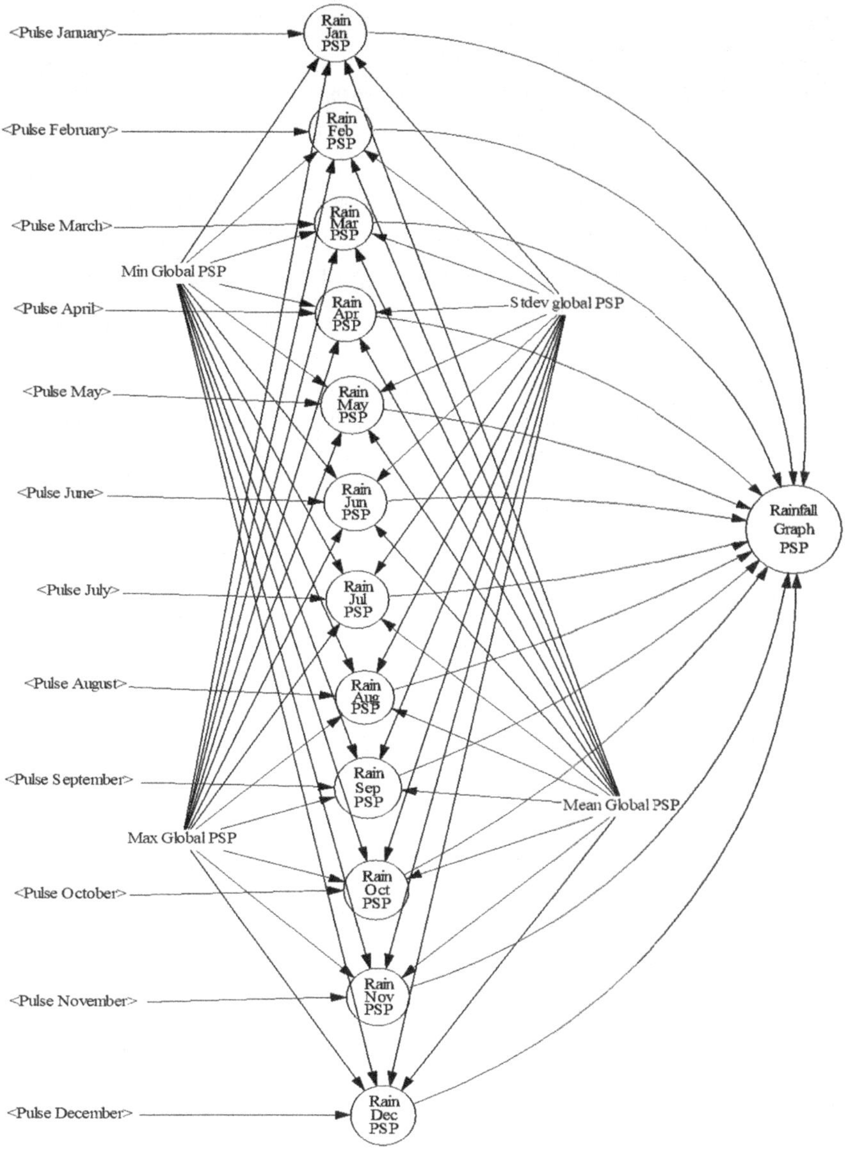

Fig. 4 SFD of rainfall categorized by month (Puspo area)

3.2 Simulation

In the simulation process, predict rainfall in four areas, which are Puspo, Sumber, Tosari, and Tutur. Each graph of simulation results of the area which includes the

Table 3 Equation used to generate rainfall value parameter each month

Variable	Equation
Rainfall graph PSP	= Rain Jan PSP + Rain Feb PSP + ... + Rain Nov PSP + Rain Dec PSP Units: mm
Rain Jan PSP	= IF THEN ELSE (Pulse January = 1, RANDOM NORMAL (4.59182*Min Global PSP, 17.1018*Max Global PSP, 10.632*Mean Global PSP, 6.83306*Stdev global PSP, 0), 0) Units: mm
...	...
Rain Dec PSP	= IF THEN ELSE(Pulse December = 1, RANDOM NORMAL (6.11364*Min Global PSP, 23.2873*Max Global PSP, 13.9725*Mean Global PSP, 9.02957*Stdev global PSP, 0), 0) Units: mm

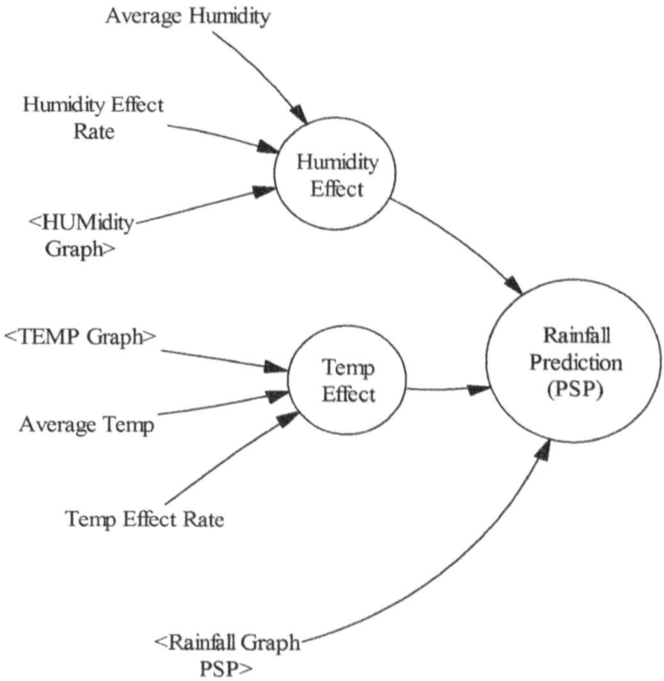

Fig. 5 SFD of rainfall prediction based on effect of temperature and humidity

Table 4 Equation used to calculate rainfall prediction based on temperature and humidity

Variable	Equation
Rainfall prediction (PSP)	= Rainfall Graph PSP − (((Temp Effect + Humidity Effect)/2)/100* Rainfall Graph PSP) Units: mm
Humidity effect	= IF THEN ELSE((HUMidity Graph − Average Humidity > 0), (Humidity Effect Rate*POWER ((1 + (Humidity Effect Rate/100)), HUMidity Graph − Average Humidity)), (Humidity Effect Rate*POWER((1 − (Humidity Effect Rate/100)), Average Humidity − HUMidity Graph))* − 1) Units: percent
Temp effect	= IF THEN ELSE((TEMP Graph − Average Temp > 0), (Temp Effect Rate*POWER((1 + (Temp Effect Rate/100)), TEMP Graph − Average Temp)), (Temp Effect Rate*POWER ((1 − (Temp Effect Rate/100)), Average Temp − TEMP Graph)) * −1) Units: percent
Average humidity	= 71.9 Units: percent
Humidity effect rate	= 4.3 Units: percent
Average temp	= 26 Units: °C
Temp effect rate	= 6 Units: percent

effects of temperature and humidity is shown in Figs. 6, 7, 8, 9 in which the simulation results look similar with the actual data.

4 Result and Discussion

Predictions model of rainfall with systems dynamics approach must be validated by the mean comparison ($E1$) and variance comparison ($E2$). The model is valid if $E1 \leq 5\%$ and $E2 \leq 30\%$. The simulation results using system dynamics without considering the effect of temperature and humidity are shown in Table 5.

However, by considering the factor of temperature and humidity on the simulation process, the RMSE is decreased up to 0.57%. Validation of predicted results is shown in Table 6.

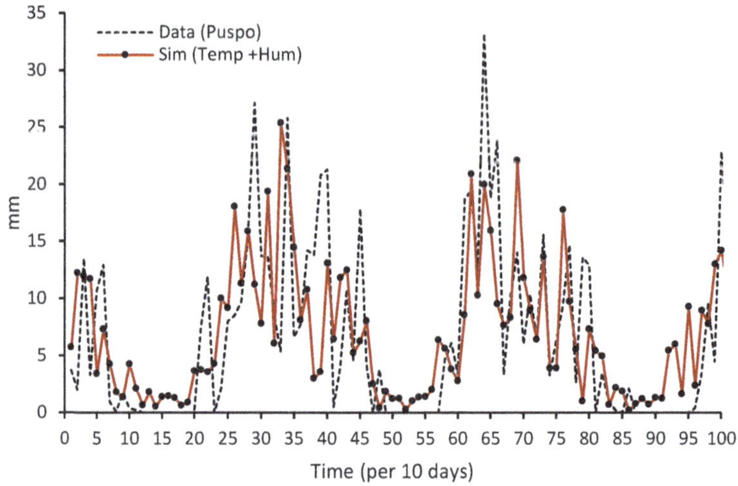

Fig. 6 Simulation graph of 100 data in Puspo

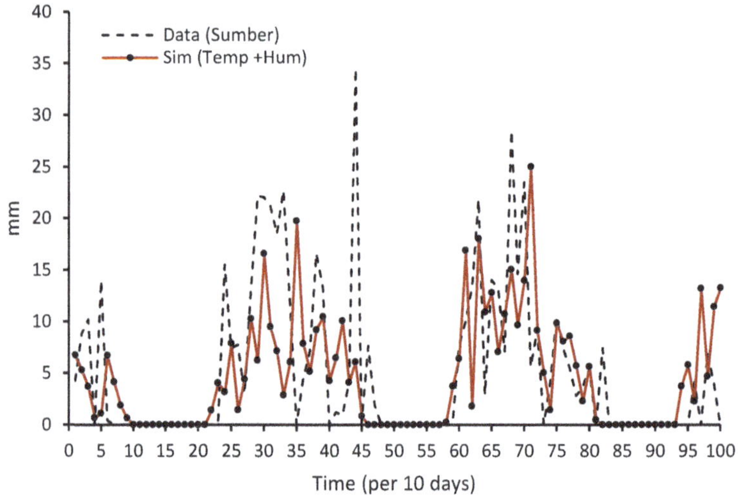

Fig. 7 Simulation graph of 100 data in Sumber

RMSE of simulation results using system dynamics is quite small. The RMSE is better when compared with previous studies using Tsukamoto FIS [3] and Hybrid Tsukamoto FIS with genetic algorithm [4] methods which only uses time series data only. The RMSE of the predicted results using system dynamics compared to a previous study are shown in Table 7.

As explained earlier, the system dynamics can predict rainfall for some period in the future. Rainfall for the year 2016 will be predicted using system dynamics with

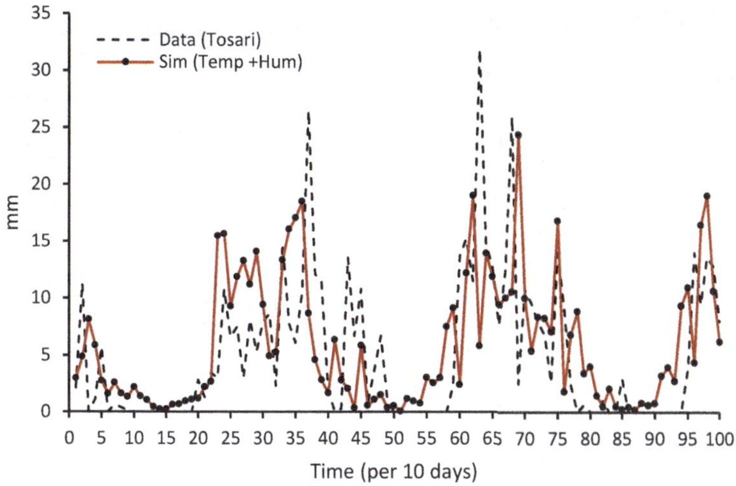

Fig. 8 Simulation graph of 100 data in Tosari

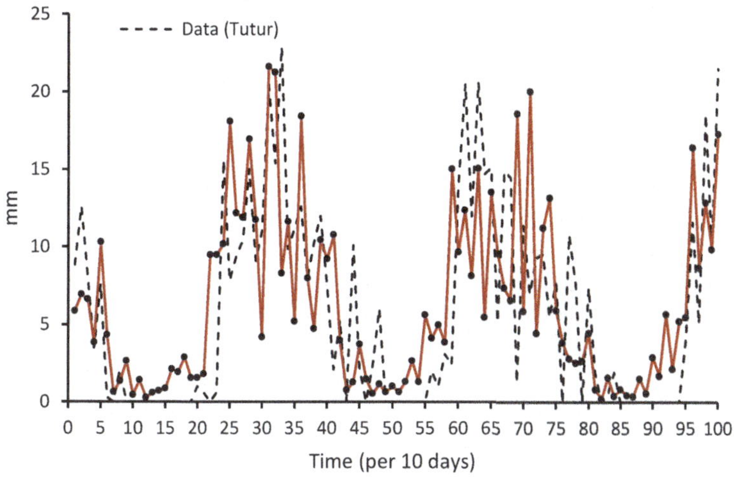

Fig. 9 Simulation graph of 100 data in Tutur

Table 5 Table captions should be placed above the tables

Region	$E1$ (%)	$E2$ (%)	RMSE
Puspo	0.8	28.4	6.7817
Sumber	4.92	28.08	7.0754
Tosari	1.0	23.3	6.4585
Tutur	3.0	25.9	6.6161

Table 6 Model validation of rainfall prediction with the temperature and humidity factors

Region	$E1$ (%)	$E2$ (%)	RMSE
Puspo	2.9	26.3	6.7672
Sumber	2.77	29.9	7.0756
Tosari	1.6	26.7	6.4219
Tutur	1.0	28.4	6.5810

Table 7 RMSE comparisons

Region	System dynamics	Hybrid FIS Tsukamoto and GA	FIS Tsukamoto
Puspo	6.7672	7.30	8.95
Sumber	7.0756	7.09	9.64
Tosari	6.4219	6.78	8.81
Tutur	6.5810	6.63	8.64

a dataset of every ten days. Rainfall prediction results in mm for the year of 2016 are shown in Table 8.

5 Conclusion

Defining the problem and model development using system dynamics can be used to predict rainfall in the Tengger, Indonesia. The model is created with input rainfall data from January 2005 until December 2014, by considering the data of temperature and humidity. Based on the prediction results using system dynamics approach, the RMSE values are smaller when compared to previous study that only uses rainfall data on previous years or time series data. This indicates that other variables such as temperature and humidity can affect the outcome of rainfall prediction. Further research can be considering several other variables that affect the rainfall which is the difference in air pressure, solar radiation, and evaporation [2].

Table 8 Rainfall prediction (mm) for 2016 using system dynamics

Month	Days	Puspo	Sumber	Tosari	Tutur
Jan	10	3.60666	8.92034	16.8427	13.28169
	20	17.98568	11.3789	6.45772	11.32804
	30	13.50693	11.0151	22.01581	10.4066
Feb	10	7.42941	3.88459	12.2626	23.83444
	20	7.90138	2.56354	10.83016	10.0027
	30	12.73383	0.94391	13.8059	14.9461
Mar	10	9.23171	8.63715	10.44825	9.49454
	20	14.70633	1.0568	9.94566	18.81341
	30	15.51178	2.19569	11.65708	3.51294
Apr	10	7.49767	1.00392	3.93915	6.58243
	20	12.31449	2.22554	13.06848	3.7001
	30	3.35869	0.87561	7.56916	2.98633
May	10	1.52366	3.54466	4.65655	2.8309
	20	5.40357	8.61088	9.49735	5.03262
	30	0.22282	10.40515	2.14106	1.45282
Jun	10	5.78393	5.30171	4.15443	0.73614
	20	0.51933	3.77267	1.89495	1.14517
	30	0.44696	0.4622	0.40581	0.57536
Jul	10	0.65866	0.29482	1.43409	0.83774
	20	0.70791	0.01222	1.6357	0.31062
	30	0.49766	0.0457	0.19722	0.71329
Aug	10	0.54773	0.4262	0.33968	0.63686
	20	1.35257	0.02915	0.18808	0.97364
	30	1.22616	1.49125	0.69283	1.0302
Sep	10	1.0805	0.4033	0.81633	2.80424
	20	1.21492	0.57883	1.30877	2.15894
	30	2.63948	0.74233	1.99675	3.06185
Oct	10	3.69109	0.41304	3.25011	2.09783
	20	1.6483	0.52985	2.59327	3.48296
	30	6.34198	3.22065	15.93162	4.58543
Nov	10	6.10176	2.28923	2.74183	15.82937
	20	7.08301	3.93986	12.62034	3.73336
	30	17.77598	5.28244	15.94338	16.33969
Dec	10	21.00568	1.15372	22.98655	15.55262
	20	19.89133	14.21413	24.27358	16.33601
	30	2.5996	11.85771	11.4881	8.94443

Acknowledgements This research was supported by the local Meteorological and Geophysics Agency Tengger, East Java and Faculty of Computer Science, Brawijaya University.

References

1. Iriany, A., Mahmudy, W.F., Sulistyono, A.D., Nisak, S.C.: GSTAR-SUR model for rainfall forecasting in tengger region, east Java. In: 1st International Conference Pure Applied Research University Muhammadiyah Malang, 21–22 August, vol. 1, pp. 1–8. (2015)
2. Fadli, M.: Prediksi (Nowcasting) Curah Hujan Pagi dan Siang Hari di Wilayah Jakarta dengan Model ANFIS (Prediction (Nowcasting) Rainfall Morning and Daylight in Jakarta Region with ANFIS Model). Megasains **3**(2), 61–76 (2012)
3. Wahyuni, I., Mahmudy, W.F., Iriany, A.: Rainfall prediction in tengger region-indonesia using tsukamoto fuzzy inference system. In: 1th International Conference Information Technology Information Systems Electrical Engineering, pp. 1–11. (2016)
4. Wahyuni, I., Mahmudy, W.F.: Rainfall prediction in tengger-indonesia using hybrid tsukamoto FIS and genetic algorithm. J. ICT Res. Appl. 1–8 (2016)
5. Adipraja, P.F.E.: *Manajemen Aset Jaringan Distribusi Energi Listrik untuk Meningkatkan Keandalan Jaringan (Studi Kasus PLN Pamekasan)* (2015)
6. Dyson, B., Chang, N.: Forecasting municipal solid waste generation in a fast-growing urban region with system dynamics modeling. Waste Manag **25**, 669–679 (2005)
7. Retno, D., Saputro, S., Wigena, A.H., Djuraidah, A.: Model Vektor Autoregressive Untuk Peramalan Curah Hujan di Indramayu (Autoregressive Vector Model For Rainfall Forecasting in Indramayu). Forum Stat. dan Komputasi **16**(2), 7–11 (2011)
8. Siswanti, K.Y.: Model Fungsi Transfer Multivariant dan Aplikasinya untuk Meramalkan Curah Hujan di Kota Yogyakarta (Multivariant Transfer Function Model and Its Application to Predict Rainfall in Yogyakarta City). Fak. Mat. dan Ilmu Pengetah. Alam—Univ. Negeri Yogyakarta, pp. 1–180. (2011)
9. Adipraja, P.F.E., Suryani, E., Hendrawan, R.A.: Manajemen Aset Jaringan Distribusi Energi Listrik: Sebuah Pendekatan Sistem Dinamik. SISFO—J. Sist. Inf. **142**, 1–9 (2015)
10. Richardson, G.P.: Encyclopedia of operations research and management science: system dynamics (2013)
11. Sterman, J.: Business dynamics—systems thinking and modeling for a complex world. J. Oper. Res. Soc. **53**(4), 472–473 (2015)
12. Barlas, Y.: Formal aspects of model validity and validation in system dynamics. Syst. Dyn. Rev. **12**(3), 183–210 (1996)
13. Trenberth, K.E.: The impact of climate change and variability on heavy precipitation, floods, and droughts. Encycl. Hydrol. Sci. 1–11, (2005)
14. NOAA: "NOAA Satellite and Information Service July, vol. 28, [online] (2016). Available. http://www7.ncdc.noaa.gov/CDO/dataproduct

Adequacy Assessment of Grid-Connected Nanogrid

Danang Wijayanto, Edi Leksono, and Augie Widyotriatmo

Abstract Microgrid, an autonomous power distribution system that utilizes distributed generation, offers advantages by means of increasing of reliability, efficiency, etc. Currently, the microgrid implementation extends to a smaller scale called nanogrid. Increasing number of microgrids and nanogrids raises the need for a measurement tool that can be used to assess the performance level of a microgrid system so that it can be used to compare one system with another, to measure the ups and downs of a system's performance over a time period, and to use the evaluation result as the base of further improvement. This paper offers a method for evaluating the reliability of a nanogrid system using a Markov model. The reliability indicator selected in this study is the loss of energy expectation (LOLE), which is considered to be more suitable for this purpose. As the validation of this model, the reliability evaluation of a nanogrid in the energy management laboratory, Institut Teknologi Bandung (ITB) is reported.

Keywords Microgrid · Nanogrid · Reliability · Loss of energy expectation (LOLE) · Markov model

1 Introduction

Microgrid is an autonomous local power system that involves the utilization of distributed energy generation using small-sized power plant located close to the consumer. The main components of microgrids are distributed generators, energy storage systems, and power electronics devices which operate to supply power to the loads. The microgrid is equipped with a network of energy distribution lines, communication lines, and a control center that monitors and controls the operation of the system. A lot of benefits can be obtained by implementing the concept of microgrid such as cutting transmission costs, utilization of renewable energy sources,

D. Wijayanto (✉) · E. Leksono · A. Widyotriatmo
Bandung Institute of Technology, Jawa Barat, Bandung 40116, Indonesia
e-mail: danang.pgrad@gmail.com

E. Joelianto et al. (eds.), *Cyber Physical, Computer and Automation System*, Advances in Intelligent Systems and Computing 1291,
https://doi.org/10.1007/978-981-33-4062-6_18

increasing system reliability, decreasing power costs, and so forth. The installed microgrid capacity varies from a few kW to tens of MW [1].

There are two modes of microgrid operation namely the "island mode" and the "connected mode" [2, 3]. In the "island mode," microgrids operate independently, meaning that it is completely separated from the main power system. It is called the "island mode" because, in this mode, the microgrid is seen to be similar as a small island separated from the mainland in terms of the main power system unit. Figure 1 illustrates a microgrid system consisted of some distributed generators, a single lumped loads, a control center governing the operation of the system. The microgrid is connected to the main grid via point of common coupling (PCC). The reliability of a microgrid can be enhanced by connecting the microgrid with the primary utilities and the energy storage systems [2, 3].

In the "connected mode," the microgrid is connected to the main power network. The gate or interconnection point between the microgrid and the other grid is often referred as the PCC [3]. From the main power system side, the microgrid in a connected state is seen as a special entity that can act as a consumer power or load, and, on the other hand, it can also serve as an electrical power producer. The term used for this phenomena is the active distribution network [1, 2]. It is said to be active because the microgrid has its own generator. Thus, it is possible to export the excess power to the main utility if the power generated exceeds its internal need.

The microgrid configuration has more variations than that of conventional power systems [1–3]. This is due to several new things that exist in the microgrids and not existed yet in the conventional power systems. These include the decentralization of generation with the use of distributed energy sources, the intensification of information and communication technology use, and the decentralization of control

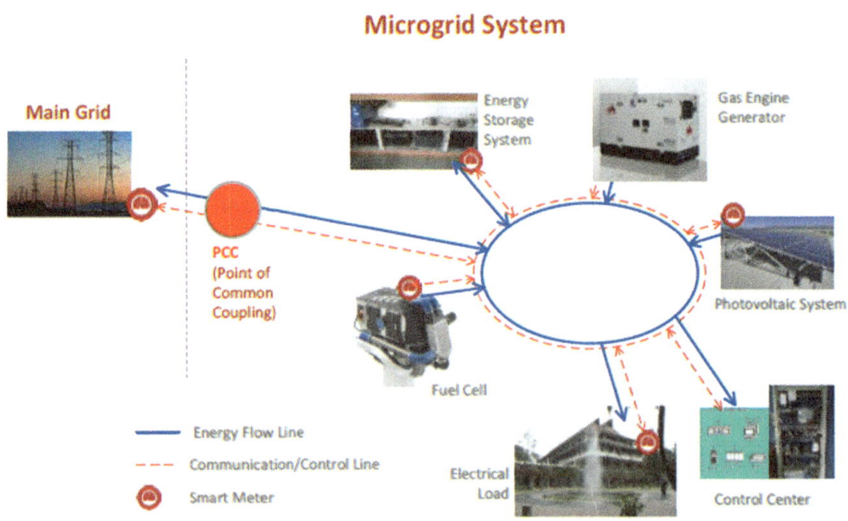

Fig. 1 Illustration of microgrid system

with the island mode. These three things provide various beneficial consequences in terms of improving the efficiency and reliability of the system. However, there is another consequence that must be embraced, which is the increasing complexity of the system. This complexity challenges the researcher to overcome and also the opportunity research in this field is still very open.

Experience in conventional power systems recommends the importance of reliability analysis to reduce customer failure. High failure rates provide a very detrimental effect on both consumers and institutions of power providers such as decreasing of productivity, declining of sales value, wasting of resources, and so on [1, 2, 4, 5].

Microgrid performance can be seen from various points of view such as efficiency, reliability, environment, and so on. This paper assesses the performance only from one point of view namely the availability of energy services to power system users. This performance was chosen because the availability of energy services is the most important aspect of a power system. In order for a microgrid to provide the expected benefits, the assurance of the system is required to be performed well and be capable to supply power as required throughout its operational life [1, 4, 6]. Supply adequacy indicators that can be used for this purpose are loss of load expectation (LOLE), loss of load probability (LOLP), loss of load frequency (LOLF), and loss of load duration (LOLD).

Reliability analysis methods for conventional energy systems today especially the adequacy analysis can be said to be very mature. In contrast, the reliability analysis of microgrid systems is relatively new and not many papers have discussed this topic. Research related to reliability and microgrid adequacy analysis is one of the biggest challenges faced in the world of electrical power today [1]. This is due to various characteristics of microgrids that are unique and somewhat different from the conventional power systems. New innovations are needed to get some reliability analysis method that can be accepted by various power system stakeholders.

References [4, 6] offer a model of relational analysis on microgrid systems by modifying the Markov model commonly used for assessing the conventional power systems. This paper essentially a follow up of their research by implementing the model on a small microgrid or a nanogrid system. Burmester in [7] defines the nanogrid as follows: "Nanogrid is a power distribution system in a building unit with the ability to connect or disconnect to other power distribution systems through a gate. Nanogrid consists of one or several power plants, distribution systems, and with a control center for local use, with a choice of likelihood equipped with a power storage system. Nanogrid capacity is generally quite small, i.e., no more than 10 kW."

The future potential of future nanogrid systems is so promising so that the need for a tool that can be used to measure its performance levels is unquestionable. From the ability of nanogrid to utilize renewable small-scale energy sources that have intermittent characteristics, its declining investment prices, its ability to operate alongside the main electrical grid, and its potential to aggregate into microgrid systems, the existence of future nanogrids is very promising. Moreover, for developing countries whose reliability of its electrical network is not very good, nanogrid is very useful because it can increase the level of reliability.

Adequacy assessment of nanogrid is needed because it can be used as a tool to carry out objective comparisons between a nanogrid system with another one or to measure its performance degradation or its performance enhancement. Unfortunately, the research on the adequacy assessment of microgrids even more on nanogrid is a few in this time. This paper is intended as an effort to give a very small contribution to fill this gap. This study offers a method of evaluating the performance of nanogrid reliability. To validate the methods offered, the existing nanogrid system in the energy management laboratory in ITB's engineering physics department is evaluated.

2 Nanogrid

Nanogrid is a microgrid in a smaller scale. Nanogrid research is relatively new field but currently it is getting a lot of attention that can be seen from the increasing number of papers that discuss this topic [7]. Similar to microgrids, nanogrid is a power distribution system that serves the power needs of a small, autonomous (self-regulating) community and generally involves one or more distributed plants using renewable energy sources. Figures 2 and 3 show the block diagram of a nanogrid and a microgrid formed from several nanogrids, respectively.

The main component of nanogrid in this study consists of a distributed power plant in the form of a photovoltaic system, battery energy storage system, load, and control system that governs the operation of the system. The nanogrid is connected to the main grid through a common point called PCC. In this system, the PCC task is performed by a smart converter. Figure 4 shows the nanogrid configuration in the ITB Energy Management Laboratory that becomes the object of this study.

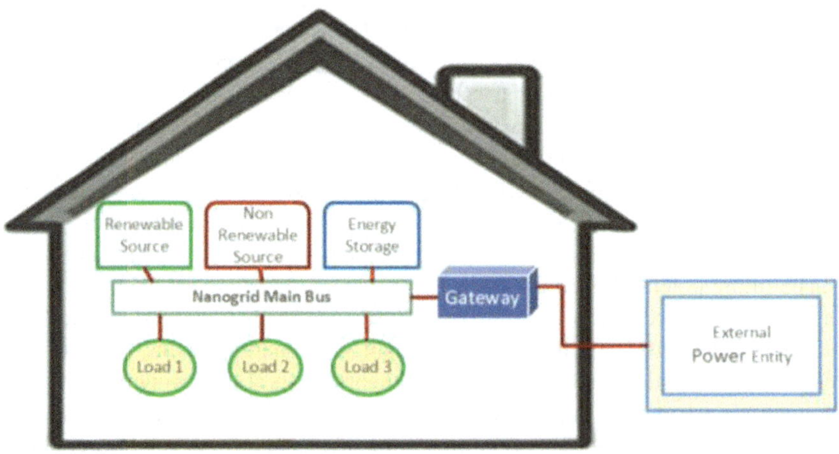

Fig. 2 Nanogrid is a small-scale microgrid that serves very limited consumers (one building)

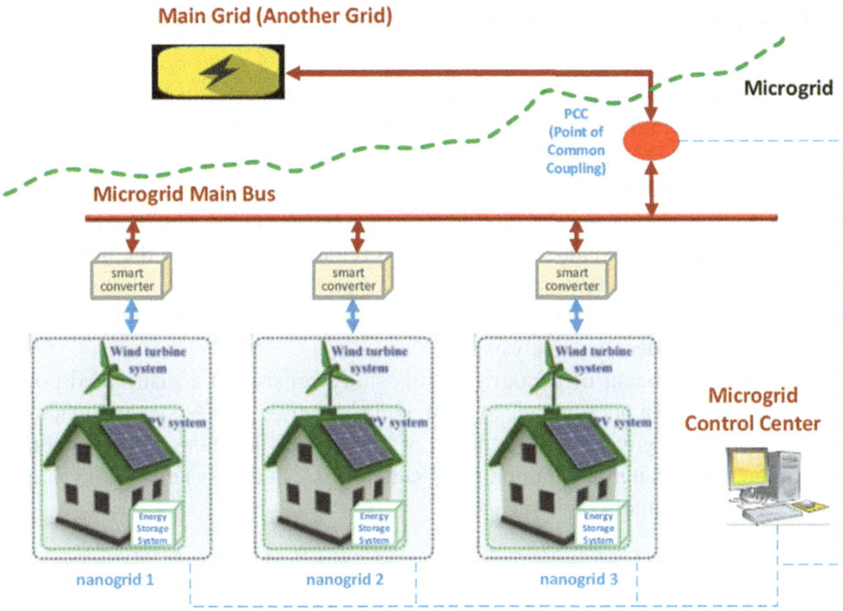

Fig. 3 Several nanogrids are integrated into a microgrid system to improve its overall performance

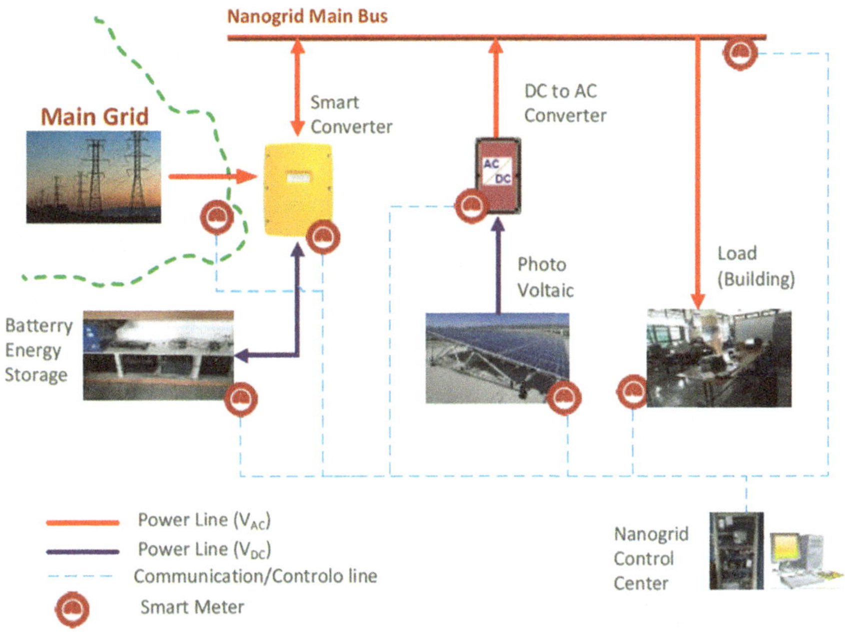

Fig. 4 Schematic diagram of nanogrid at the energy management Laboratory, engineering physics department, Institut Teknologi Bandung

3 Reliability Analysis of Power System

Figure 5 shows the various points of view to analyze the performance of an electric power system. The reliability analysis of electric power system can be grouped into three point of views, which are the adequacy's point of view, the security-and-safety's point of view and the quality's point of view. Adequacy analysis evaluates the ability of the system to supply energy to consumers according to the needs. Any system failure to meet these demands will degrade its reliability value. On the other hand, the safety and security analysis is a study that responses to the system transition and at the same time how to overcome [5]. This study only discusses the first point of view which is the adequacy analysis.

The main component of nanogrid in this study consists of a distributed power plant in the form of a photovoltaic system, battery energy storage system, load, and control system that governs the operation of the system. The nanogrid is connected to the main grid through a common point called PCC. In this system, the PCC task is performed by a smart converter.

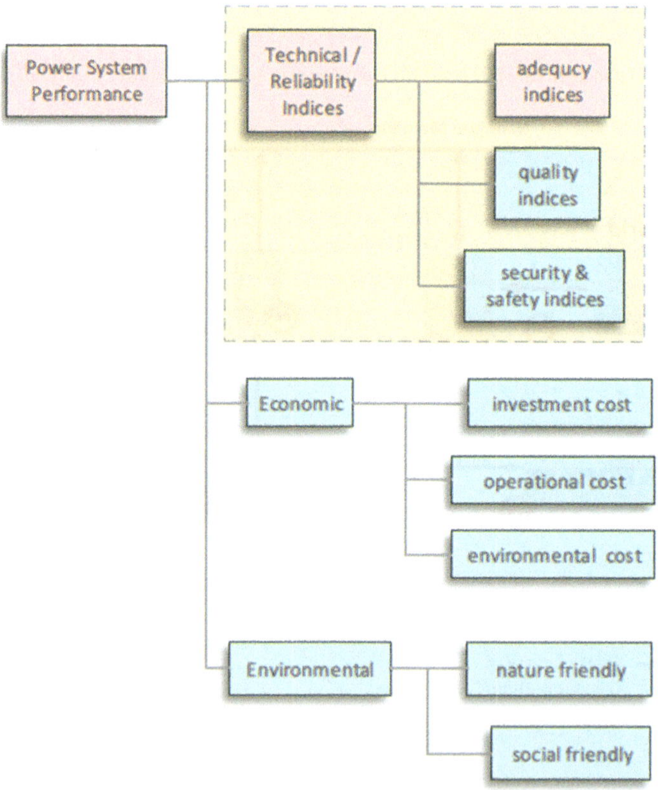

Fig. 5 Miscellaneous aspect of power system performance

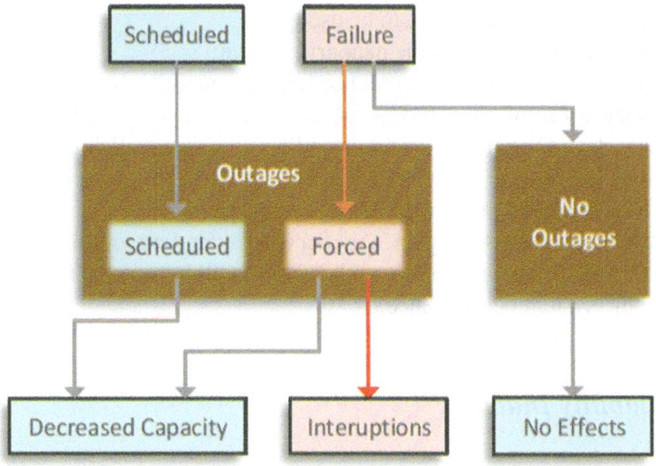

Fig. 6 Block diagram showing the relationship between failure, outage, and interruption

3.1 Outage and Interruption

The most influential factor to the reliability value of the power system is the outage. Outage is a situation when a major power system device does not work due to various reasons. Outage events can lead to interruptions, that is the event when the load experience loss of power supply. Outage can take a place partly or on a whole system [5, 8].

There are two kinds of outage, i.e., deliberate and accidental [5, 9]. Deliberate outage or scheduled outage is taken place during maintenance on the system component. Unintentional outages occur when a failure occurs in one component causing the equipment to fail or when an emergency occurs causing a device to be disabled to prevent further damage or accidents. This forced outage reduces the reliability value of the power system because it leads to a breakdown of the power supply to the consumer, or to an interruption [5]. Not all failures cause an outage and not all outages cause interruptions. The relationship between failure, outage, and interruption is explained by Fig. 6.

3.2 Stages of Nanogrid Adequacy Assessment

There are two common approaches to evaluate a system reliability, which are the analytical method and the simulation method [5]. The assessment method conducted in this study is classified as the analytical approach. The stages of analytic approach used in power system reliability analysis of the nanogrid are as follows:

- Determine the reliability index to be used;
- Create a reliability model of the nanogrid system;
- Collect the required data in accordance with the model used;
- If necessary, try some case studies consist of miscellaneous scenarios to compare each other. By doing this, the understanding of system reliability behaviors can be improved;
- Do analysis based on predefined models and data;
- Formulate conclusions to gain a better understanding of the systems being analyzed. A better understanding of the system will be very useful in the effort to improve the system performance.

3.3 Reliability Indicator

This paper will evaluate one aspect of nanogrid performance namely the continuity service or the availability of supply energy to the power system consumer. This performance is chosen because the availability of energy services is the most important aspect of a power system. The power supply system reliability indicator used is lost of load expectation (LOLE). LOLE measures the reliability level of a system based on the losses occurred when the system fails to supply its load demand, because of the failure on some system components [1, 5].

The provider of the power system must be able to ensure that the power generation capacity is sufficient for the duration of its operation. LOLE is one of the best reliability indicators used for this purpose. As the name implies, LOLE quantifies the magnitude of "loss of load events" in the power system. "Loss of load" is an event the magnitude of system exceeds the generation capacity. The small LOLE value indicates the good power system reliability and vice versa. LOLE estimates the duration of loss of load loss in a power system within a specified span of time (usually 1 year) by taking into account the stochastic data of the system. Figure 7 shows the three steps to calculate LOLE [5].

The equation for calculating power system LOLE during its operation period is as follows:

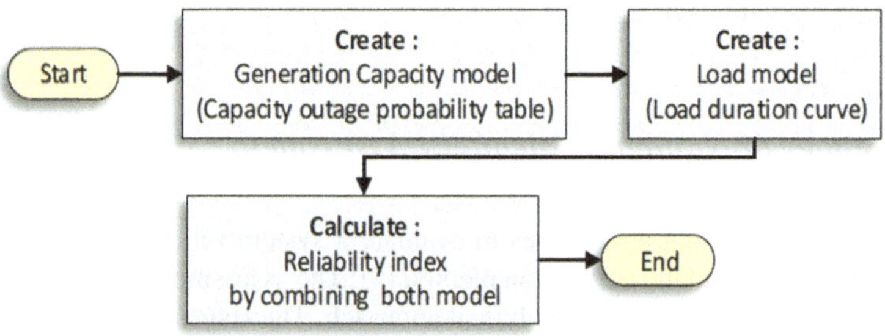

Fig. 7 LOLE calculation step

$$\text{LOLE} = \sum_{i=1}^{n} P_i(C_i - L_i) \text{ days/period} \tag{1}$$

where C_i is the available capacity on day *i*, L_i is the forecast peak load on day *i*, and P_i $(C_i - L_i)$ is the probability of loss of load on day *i*.

4 Result and Discussions

After determining the LOLE as the reliability index to be used, the next step is to set up the reliability model to be used. The common model used for this purpose is the Markov model [5]. The Markov model is used to evaluate the effect of generating units on the reliability of the power system (adequacy assessment) by taking into account the failure rate λ and repair rate μ of the components.

Overall, the nanogrid reliability can be estimated by analyzing the reliability indicators of the main components separately. By using the Markov model, the stochastic nature of failure event of the major components is analyzed. The Markov model defines the operation of a system or component into multiple states which takes into account the probability of moving from one operational status to another [5].

From the reliability point of view, the nanogrid system can be grouped into four entities of electrical power supply units which can be modeled into a Markov reliability model. Those entities are:

- Main utility;
- Smart converter;
- Battery energy storage;
- Photovoltaic system.

Markov models for main utility, smart converter, and battery energy storage are similar to each other, because these components have only two states that are up state, i.e., when the system component operates normally, and down state, i.e., when otherwise. For the photovoltaic system, there are two components that are photovoltaic and inverter and each has two possible states of up and down. Then, the total possible states for photovoltaic are four. The Markov model of both entities is shown in Figs. 8 and 9. In Fig. 8, the subscript "sub" is interchangeable with "ut" for utility, "es" for energy storage, and "cs" for smart converter. In Fig. 9, the subscript "inv" is for inverter and "pv" for photovoltaic.

Availability A_{sub} and unavailability U_{sub} for the model shown in Fig. 8 are described as:

$$A_{\text{sub}} = \frac{\mu_{\text{sub}}}{\lambda_{\text{sub}} + \mu_{\text{sub}}}, \tag{2}$$

$$U_{\text{sub}} = 1 - A_{\text{sub}} \tag{3}$$

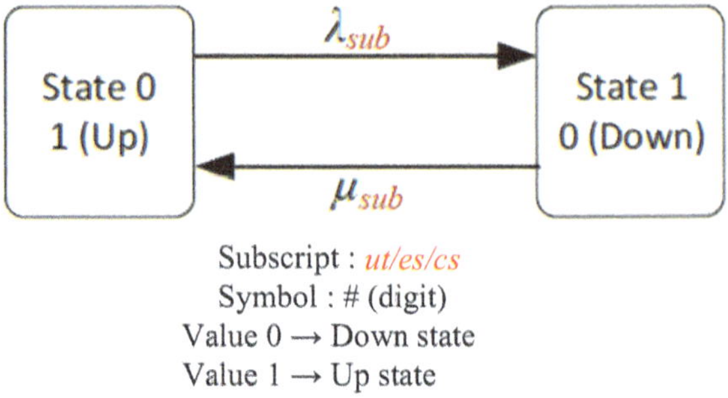

Fig. 8 Reliability model of utility, energy storage, and smart converter

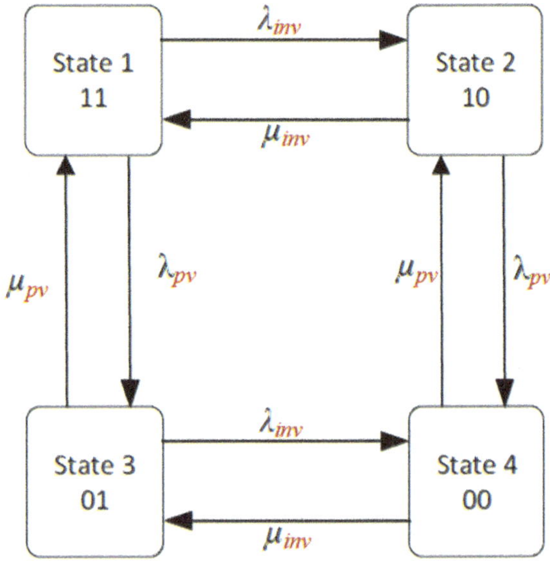

Fig. 9 Reliability model of photovoltaic system

and those for the model shown in Fig. 9 are:

$$\begin{aligned} A_{\text{pvs}} &= A_{\text{pv}} \cdot A_{\text{inv}} \\ &= \left(\frac{\mu_{\text{pv}}}{\mu_{\text{pv}} + \lambda_{\text{pv}}}\right) \cdot \left(\frac{\mu_{\text{inv}}}{\mu_{\text{inv}} + \lambda_{\text{inv}}}\right), \end{aligned} \tag{4}$$

$$U_{\text{pvs}} = 1 - A_{\text{pvs}} \tag{5}$$

where the subscript "pv" is for the photovoltaic systems including the photovoltaic and the inverter.

To gain a better understanding of the reliability behavior of the system under study, some configuration scenarios are investigated. Each scenario is analyzed and the results are compared each other. The two scenarios are:

Table 1 Technical parameter of nanogrid

Parameter	Main grid	PV	Inverter
Installed capacity (kW)	2	1	2
Nominal capacity (kW)	2	1	2
Life time	–	20 year	20 year
Efficiency (%)	99	19	99
Failure rate (f/y)	5.3	0.04	0.143
Repair rate (f/y)	73	18.25	52.143

- The system is served only by one power source, i.e., from the main utility;
- Nanogrid is served by two power sources, i.e., main utility and photovoltaic system.

The technical specifications of each of the major components of the nanogrid system are shown in Table 1. Every minute, the supervisory control and data acquisition (SCADA) system on the nanogrid system perform data acquisition of electrical power load. The "per minute data" is averaged for each one hour period, so that 8760 load data for one year is obtained. From this data, the load profile of the system in the form of load duration curve is obtained. Load duration curve is similar to a load curve but the demand data is ordered in descending order of magnitude, rather than chronologically. The load duration curve of the nanogrid is shown in Fig. 10.

In the first scenario, the stochastic behavior of total system generation is represented by the value of availability and unavailability. From Table 1, the failure rate of the utility source (λ_{ut}) is 5.3 (failure/year) and its repair rate (μ_{ut}) is 73 (repair/year).

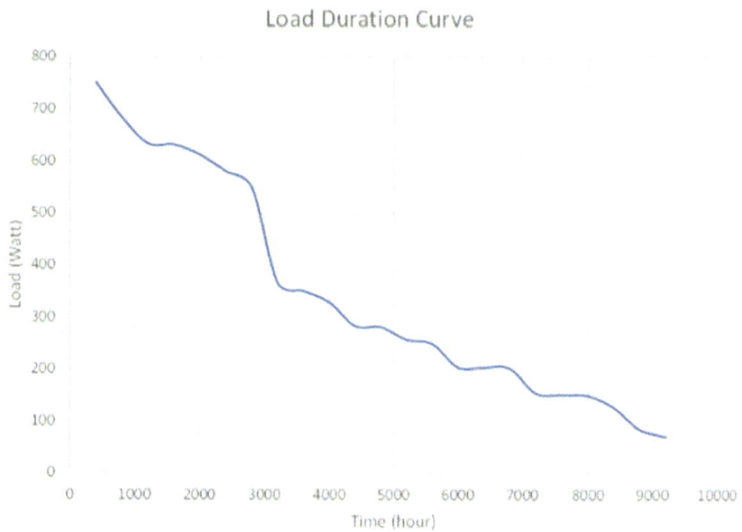

Fig. 10 Load duration curve of the nanogrid

Total calculated availability and unavailability of the generation system is $A_{ut} = 0.93$ and $U_{ut} = 0.07$, respectively. Since the power supply system comes from only one source, the LOLE system is:

$$\begin{aligned} \text{LOLE} &= \sum_{i=1}^{n} P_i(C_i - L_i) \\ &= \sum_{i=1}^{n} U_i = 0.07 \times 365 = 25.55 \frac{\text{day}}{\text{year}} \end{aligned} \tag{6}$$

From the reliability point of view, the system configuration of the second scenario can be described as in Fig. 11. Because the quantity of power source is more than one, the stochastic characteristics of total system generation are represented by using the capacity outage probability table (COPT). COPT is an electrical generation system model in the form of a table containing a list of outage capacities and its probability of occurrence.

The value of loss of load probability (LOLE) can be obtained by comparing the COPT with the system load profile. To create a COPT, we must first know the availability and unavailability values of the power sources:

- A_{ut} and U_{ut}: Availability and unavailability of the utility source has been calculated before.
- A_{pvs} and U_{pvs}: Availability and unavailability of solar generating systems can be calculated using Eq. (3).

The results obtained are as follows: $A_{ut} = 0.93$, $U_{ut} = 0.07$, $A_{pvs} = 0.98$, $U_{pvs} = 0.02$.

Table 2 shows the result of the capacity outage probability for the second scenario. Based on the load duration curve data and COPT data, the LOLE calculation is

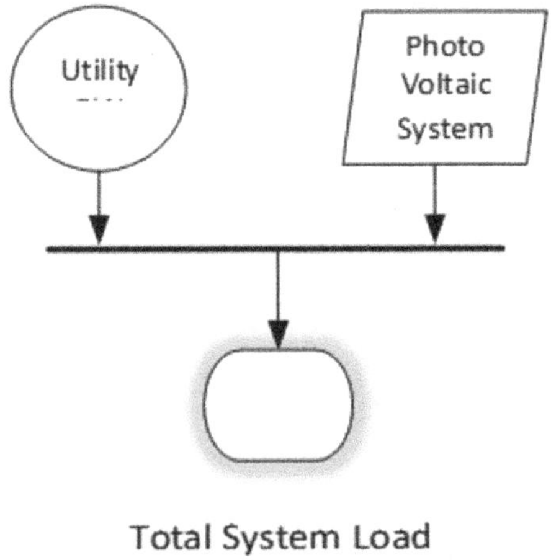

Fig. 11 Adequacy representation of case-2

Table 2 Capacity outage probability of Scenario 2

Outage capacity (W)	Individual probability	Cumulative probability
0	0.9114	1.0000
190	0.0186	0.0886
1980	0.0686	0.0700
2170	0.0014	0.0014

Table 3 Load data for LOLE calculation of Scenario 2

Load range	Time range	Duration (h)
Load $\leq$ 190 W	Hour: 6951–8760	1810
Load > 190 W	Hour: 0–6950	6950

summarized as follows (Table 3):

$$\begin{aligned}
\text{LOLE} &= \sum_{i=1}^{n} p_i(C_i - L_i)\,\frac{\text{h}}{\text{year}} \\
&= 1810.P(\text{Load} < 190) + 6950.P(\text{Load} \geq 190) \\
&= 1810.P(2170 - 189) + 6950.P(2170 - 191) \\
&= 1810.P(1981) + 6950.P(1979) \\
&= 1810.(0.0014) + 6950.P(0.0700) \\
\text{LOLE} &= 489\,\text{h/year}, \\
\text{LOLE} &= \frac{489\ \text{h/year}}{24\ \text{h/day}} 20.38 \frac{\text{days}}{\text{year}}.
\end{aligned} \tag{7}$$

The calculation results in Scenario 1 and Scenario 2 show that by the utilization of distributed generator using renewable energy sources improves the system performance significantly. In Scenario 1, the system without nanogrid has LOLE value of 25.55 day/year. After implementing the nanogrid system by utilizing photovoltaic system, the system performance increases, proven from its LOLE is reduced to 20.38 day/year.

5 Conclusion

Reliability analysis is very important in power systems. There have been many techniques developed to measure the reliability of a power system. However, there is still considerable publicity with reliability analysis of microgrid systems and more nanogrid systems because microgrids and nanogrids are relatively new technologies compared to conventional electric power systems. The reliability analysis results show that the performance of the power systems increase when utilizing photovoltaic system attached to the utility grid. Further research needs to be conducted especially on the more complex systems, e.g., the number of nanogrids is more than one, the reliability studies on the complete system including energy storage systems, and other renewable energy generations. Moreover, the planning scheme can be conducted by using the LOLE calculations to evaluate various system development alternatives to obtain an investment value.

Acknowledgements This research is supported by the domestic scholarship program of Indonesian lecturer (*Beasiswa Unggulan Dosen Indonesia—Dalam Negeri, BUDI-DN*) and by the Ministry of Research, Technology and Higher Education of the Republic of Indonesia.

References

1. Bansal, R. (ed.): Handbook of Distributed Generation, Springer (2017)
2. Parhizi, S., Lotfi, H., Khodaei, A., Bahramirad, S.: State of the art in research on microgrids: a review. IEEE Access **3**, 890–925 (2015)
3. Banerji, A., Sen, D., Bera, A.K., Ray, D., Paul, D., Bhakat, A., Biswas, S.K.: MICROGRID: a review. In: IEEE Global Humanitarian Technology Conference (GHTC-SAS), Trivandrum, India (2013)
4. Adefarati, T., Bansal, R.: Reliability and economic assessment of a microgrid power system with the integration of renewable energy resources. Appl. Energy **206**, 911–933 (2017)
5. Billinton, R., Allan, R.N.: Reliability Evaluation of Power Systems, 2nd edn. Springer, (1996)
6. Adefarati, T., Bansal, R.C.: Reliability assessment of distribution system with integration of renewable distributed generation. Appl. Energy **185**, 158–171 (2017)
7. Burmester, D., Rayudu, R., Seah, W., Akinyele, D.: A review of nanogrid topologies and technologies. Renew. Sustain. Energy Rev. **67**, 760–775 (2017)
8. Casteren, J.V.: Assessment of Interruption Costs in Electrical Power Systems using Weibull-Markov Model. Goteborg (2003)
9. Billinton, R., Allan, R.N.: Reliability Evaluation of Engineering Systems—Concept and Technique, 2nd edn, pp. 453. Springer (1992)

Electroencephalography-Based Neuromarketing Using Pegasos on Partition Membership Data

Intan Nurma Yulita, Asep Sholahuddin, Emilliano, and I Gede Eka Wiantara Putra

Abstract Neuromarketing exists as a manipulation trick to map customer perceptions and desires. Through this method, companies can market their products more effectively. This study examines it by using machine learning. The data were recorded using electroencephalography from 30 respondents. A video was used to simulate the respondents. It displays product images. Then they gave an assessment of the picture presented, whether they liked it or not. For automation, we build a system that classifies based on the Pegasos method. This method was also combined with the propositionalization approach. It aims to form partition membership data. Based on the experiments that have been carried out, the combination of these two methods has much higher performance compared to the support vector machine (SVM) method. Its accuracy reaches 93.30%. It shows that the combination of the two ways is a suitable method in the classification process, especially on the neuromarketing data used.

Keywords Electroencephalography · Neuromarketing · Pegasos · Support vector machine · Propositionalization

1 Introduction

Marketing is an essential part of every company [1]. The aim is to offer products and services from the company so that buyers are interested in using them. More and more products or services purchased will increase revenue from the company [2]. Initially, the approach was compiled from data and facts in the field based on market research. But it is not enough. The company will not use too many funds to get product data. Companies also need to observe consumer behavior toward products or

---

I. N. Yulita (✉) · A. Sholahuddin · Emilliano
Universitas Padjadjaran, Sumedang 45363, Indonesia
e-mail: intan.nurma@unpad.ac.id

I. G. E. W. Putra
Politeknik Nasional, Denpasar 17, Denpasar 80239, Indonesia

E. Joelianto et al. (eds.), *Cyber Physical, Computer and Automation System*, Advances in Intelligent Systems and Computing 1291,
https://doi.org/10.1007/978-981-33-4062-6_19

services. To realize this goal, the company maximizes the use of emotional reactions when viewing advertisements, product designs, or product recommendations from endorsements [3]. This method is called neuromarketing technique. The technique has long been developed since decades ago. It exists as a popular discipline that combines neuroscience, marketing science, and social psychology [4]. In general, it examines the effect of physiological marketing and advertising strategies on the human brain. So that in the end, this method can provide the right approach in influencing potential customers. It also provides a proper understanding of the emotional reactions of prospective customers [5]. This understanding is useful for making the buyer's appeal from the advertisements for the products being displayed. In other words, neuromarketing exists as a manipulation trick to map customer perceptions [6]. Surely this manipulation trick is useful for optimizing marketing initiatives or activities that are carried out.

At present, information technology has developed so rapidly. The development of advanced technology makes our lives more modern. As technology advances to accelerate social work, the use of technology in business has been carried out [7]. The application of technology to share business interests including marketing has been carried out [8]. The business world is rushing that almost small and large-scale companies have used information technology to provide improvements to the business services they manage.

The technique that has developed a lot in the use of technology is machine learning. It is a study of algorithms to learn something in doing certain things done by humans automatically [9]. In solving problems, it makes an accurate prediction of new conclusions from various patterns that have been studied previously [10]. The success of machine learning in signal data has been widely obtained, among others, in speech data [11], EEG [12], EMG [13], polysomnography [14]. Therefore, this study uses EEG data for neuromarketing.

Support vector machine (SVM) is one of the most popular machine learning algorithms for classification [15]. In the learning process, SVM introduces a new strategy by finding the best hyperplane in the input space [16], through an approach called structural risk minimization. The algorithm has advantages in showing excellent performance for time series predictions [17]. It works well with high-dimensional data using kernel techniques. The way does not produce accurate results when many features are not relevant; not all features are needed in the process. Feature selection works directly, reducing the number of features, and selecting features that provide information on the number of features is reduced significantly, and over-fitting issues are resolved. The SVM method provides effective performance when irrelevant features are removed. Another problem with SVM is that it is biased and cannot handle large amounts of data.

These problems can be solved by using subgradients so that Pegasos is created, namely SVM, which is integrated with the stochastic subgradient descent algorithm [18]. This algorithm functions in solving optimization problems. This method will be used as a classifier in this stud. The system also added a new process, namely propositionalization that transforms data into partition membership data [19]. The existence of this propositionalization aims to facilitate Pegasos in choosing the best hyperplane in dividing between classes.

2 Research Method

The steps in this study consists of six main stages (Fig. 1).

2.1 Data Collecting

Psychological studies on the relationship between emotions and the brain reveal substantial implications of cognitive processes in emotions [20]. Thus, this research was conducted by recording electroencephalography (EEG). EEG signals carry valuable information about the emotions felt by the participants. Then, the result is that the EEG signal is recorded using active AgCl electrodes, placed according to international standards with a 10–20 system. Electrodes are placed along the scalp. This study used 14 electrodes mounted on channels including F3, F4, F7, F8, O1, O2, P7, P8, T7, T8, AF3, AF4, FC4, FC5, FC6. This dataset involved 30 people. Each respondent was simulated via a video [21]. This video contained images of several products. Respondents were asked to rate "like" or "dislike" the product that was displayed. The assessment process was done through a focus on thinking so that the EEG can record it correctly and distinguish between these two assessment labels.

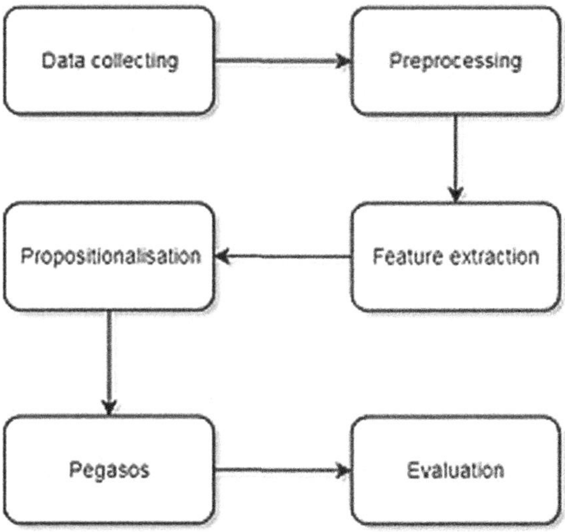

Fig. 1 Flowchart of the proposed classification in neuromarketing

EEG signal processing must be done because the visual recognition of EEG signals is challenging. The information contained in this EEG signal cannot be directly observed because the EEG signal has a weak frequency and amplitude. This study uses a notch filter and a bandpass filter to process the initial EEG data. This stage involves a bandpass filter and a notch filter. Bandpass filter is a filter that passes frequency signals within a specific frequency range. Only signals that are between the lower bound to the upper limit frequency are allowed to pass. This filter rejects or weakens the frequency signal that is outside the specified range. This filter is different from the notch filter which rejects signal components that are on a specific frequency [14]. This research is based on a second-order bandpass filter. Notch filter used $w_0 = 50 * 2/\text{sampling rate}$ and $bw = w_0/30$.

2.2 *Feature Extraction*

The next step is feature extraction, which aims to obtain feature differences between one type of EEG signal and another by interpreting signals. This process is related to the quantization of the signal essence characteristics into a group of corresponding feature values. Wavelet is one of the feature extraction methods commonly used in signals. Wavelet can analyze single and multidimensional signals, especially if these signals have different information at each time [22]. Wavelet representation is multi-scale of signal decomposition, which can be considered a tree. Each level stores a signal projection into its base function at a specific resolution. Time–frequency representation with this wavelet is done by filtering the signal with a pair of filters that will cut the frequency domain to the middle. The basic principle of DWT is a way to get the time and scale representation of a signal using digital filter techniques and subsampling operations. In this study, the EEG signal used has low frequency and amplitude characteristics. So for this case, discrete wavelet transform (DWT) is a suitable method in reducing the EEG signal.

2.3 *Propositionalization*

Propositional learning is the process of transforming the attributes of a dataset to be transformed into other attributes [19]. This transformation is done by forming a decision tree. In this study, trees were created based on the C4.5 algorithm. The leaves of the tree are the labels that the dataset presents. Propositionalization is carried out in complex cases so that the decision tree has several leaves that represent the same label. This trajectory from root to leaf is used to produce new attributes. The final result is the transformation of the original attributes into attributes derived from the results of the decision tree.

2.4 Pegasos

Support vector machine (SVM) is one of the supervised learning techniques for classification in machine learning. Unlike the neural network strategy that tries to find hyperplane separators between classes, SVM tries to find the best hyperplane in the input space. The basic principle of SVM is a linear classifier and subsequently developed to work on nonlinear problems by incorporating the concept of kernel tricks in high-dimensional workspaces. Currently, SVM has been successfully applied to real-world problems (real-world problems), and in general, provides better solutions than conventional methods such as artificial neural networks [23].

The concept of SVM can be explained simply as an attempt to find the best hyperplane that functions as a separator of two classes in the input space. Hyperplane in vector space with dimension d is affine subspace with dimension d-1, which divides the vector space into two parts, each of which corresponds to a different class [23]. The best hyperplane separator between the two categories can be found by measuring the margin of the hyperplane and finding its maximum point. Margin is the distance between the hyperplane and the closest pattern of each class. The closest model is called a support vector.

SVM transforms the data in the input space into a higher-dimensional space (feature space), and optimization is carried out on the new vector space. It distinguishes SVM from the solution method from pattern recognition in general, which performs parameter optimization on the results of transformations that are lower in dimension than the input space dimensions. But SVM is challenging to use in large-scale problems. In this case, it is intended by the number of samples processed. But SVM is challenging to use in widespread issues. SVM encounters optimization problems for the amount of sample data processed. The solution to this problem is Pegasos. Pegasos applies stochastic subgradient descent to optimization problems from SVM. Pegasos has a more straightforward and more effective process so that it will not experience problems with the increasing size of the dataset.

The Pegasos algorithm is as follows. The input of the algorithm is a dataset S, λ, several iterations T. In the first step, w_1 is assigned to vector zero. Select sample data $(x_{\text{it}}, y_{\text{it}})$ randomly, where it $\in \{1 \ldots n)$. N is the amount of data in the S dataset. Calculate y_t_t based on Eq. (1).

$$V_t = \frac{1}{\lambda_t} \quad (1)$$

For iteration 1 until T, apply that if $y_{\_\text{it}}\ y_{\text{it}}\langle w_t, y_{\text{it}}\rangle < 1$ then use Eq. (2). If not, we apply Eq. (3) to update w.

$$w_{t+1} = (1 - \gamma_t \lambda) w_t + \gamma_t y_{\text{it}} x_{\text{it}} \quad (2)$$

$$w_{+1t} = (1 - \gamma_t \lambda) w_t \quad (3)$$

2.5 Evaluation

The testing process is carried out by a cross-validation mechanism. It is a random sample sampling technique that guarantees that the total number of observed data is the same as the amount of training data and only once in the testing data. The typical standard to obtain the best error estimate is 10. The data are randomly divided into 10 parts with the same ratio then the error rate is calculated section by section, then the error rate of 10 pieces. This study also applied 10 cross-validation. The performance measurement of this study uses confusion matrix and accuracy.

3 Results and Analysis

The experiment was carried out by developing SVM and Pegasos. Furthermore, these two methods were combined with partition membership (PM) data using propositionalization.

3.1 The Performance of Base Classifiers

Base classifiers in this research were SVM and Pegasos. Tables 1 and 2 show the matrix confusion of both methods. The SVM shows failure in classifying "like" labels. None of the data with this label was successfully classified. All data by SVM were classified as "dislike" labels. Unlike the case with Pegasos, some "like" labels were classified correctly. It shows the advantages of Pegasos over SVM. But the weakness is seen on the "dislike" labels. If on SVM, all of these data were successfully classified correctly. But on Pegasos, there were data that still fail to be classified.

Table 1 Confusion matrix of SVM

Matrix	Predicted: dislike	Predicted: like
Actual: dislike	584	0
Actual: like	461	0

Table 2 Confusion matrix of Pegasos

Matrix	Predicted: dislike	Predicted: like
Actual: dislike	416	168
Actual: like	257	204

3.2 *The Performance of Combination of Base Classifiers with the Propositionalization*

In this process, the base classifiers were combined with propositionalization. SVM successfully classified all data with a "dislike" label. But it still failed to classify data with the "like" label. This performance was the same as SVM before being combined with propositionalization. It indicates that propositionalization did not influence SVM performance.

On the other hand, PM-Pegasos provided high performance. Of all the data with the "dislike" label, only 30 data failed to be classified. Such is the case with the "like" label. 40 data were unable to be classified as "like" labels. The amount of data that failed to be classified reached less than 10%. As shown in Table 5, the propositionalization had an essential influence on Pegasos. The Pegasos accuracy was 30% higher than without it.

4 Conclusion

This research implemented machine learning for neuromarketing. The mechanism was tested using data from 30 respondents. The classification used SVM and Pegasos methods. The performance of Pegasos was better than SVM, as shown in Tables 1 and 2. But overall, the accuracy operated by both was still small. The solution was to use propositionalization. The existence of propositionalization did not affect SVM. It can be seen in Table 3. The SVM even failed to classify the "like" label. Table 4 shows that only a small portion of data was unable to be classified by a combination of the propositionalization and Pegasos. The accuracy of this combination is shown in Table 5 reaching 93.30%. This high accuracy indicates that the combination of the two has succeeded in classifying neuromarketing data correctly.

Table 3 Confusion matrix of PM-SVM

Matrix	Predicted: dislike	Predicted: like
Actual: dislike	584	0
Actual: like	461	0

Table 4 Confusion matrix of PM-Pegasos

Matrix	Predicted: dislike	Predicted: like
Actual: dislike	554	30
Actual: like	40	421

Table 5 Accuracy

No.	Methods	Accuracy (%)
1	SVM	55.89
2	Pegasos	59.33
3	PM-SVM	55.89
4	PM-Pegasos	93.30

References

1. Menon, A., Bharadwaj, S.G., Adidam, P.T., Edison, S.W.: Effective marketing strategy—making: antecedents and consequences. In: Proceedings of the 1997 Academy of Marketing Science (AMS) Annual Conference, pp. 224–224. Springer, Cham (2015)
2. Olson, E.M., Slater, S.F., Hult, G.T.M., Olson, K.M.: The application of human resource management policies within the marketing organization: the impact on business and marketing strategy implementation. Ind. Mark. Manage. **69**, 62–73 (2018)
3. Pappas, N.: Marketing strategies, perceived risks, and consumer trust in online buying behavior. J. Retail. Consum. Serv. **29**, 92–103 (2016)
4. Stanton, S.J., Sinnott-Armstrong, W., Huettel, S.A.: Neuromarketing: ethical implications of its use and potential misuse. J. Bus. Ethics **144**(4), 799–811 (2017)
5. Hsu, M.: Neuromarketing: inside the mind of the consumer. Calif. Manage. Rev. **59**(4), 5–22 (2017)
6. Meckl-Sloan, C.: Neuroeconomics and neuromarketing. Cell **650**, 218–8214 (2015)
7. Linton, J.D., Solomon, G.T.: Technology, innovation, entrepreneurship, and the small business—technology and innovation in small business. J. Small Bus. Manage. **55**(2), 196–199 (2017)
8. Anzoategui, D., Comin, D., Gertler, M., Martinez, J.: Endogenous technology adoption and R&D as sources of business cycle persistence. Am. Econ. J. Macroecon. **11**(3), 67–110 (2019)
9. Yulita, I.N., Purwani, S., Rosadi, R., Awangga, R.M.: A quantization of deep belief networks for long short-term memory in sleep stage detection. In: IEEE 2017 International Conference on Advanced Informatics, Concepts, Theory, and Applications (ICAICTA), pp. 1–5 (2017)
10. Yulita, I.N., Fanany, M.I., Arymurthy, A.M.: Fuzzy clustering and bidirectional long short-term memory for sleep stages classification. In: IEEE 2017 International Conference on Soft Computing, Intelligent System and Information Technology (ICSIIT), pp. 11–16 (2017)
11. Yulita, I.N., Hidayat, A., Abdullah, A.S., Awangga, R.M.: Feature extraction analysis for hidden Markov models in Sundanese speech recognition. Telkomnika **16**(5) (2018)
12. Djamal, E.C., Gustiawan, D.P., Djajasasmita, D.: Significant variables extraction of post-stroke EEG signal using wavelet and SOM kohonen. Telkomnika **17**(3) (2019)
13. Too, J., Abdullah, A.R., Saad, N.M., Ali, N.M., Zawawi, T.T.: Application of Gabor transform in the classification of myoelectric signal. Telkomnika **17**(2), 873–881 (2019)
14. Yulita, I.N., Fanany, M.I., Arymurthy, A.M.: Fast convolutional method for automatic sleep stage classification. Healthc. Inf. Res. **24**(3), 170–178 (2018)
15. Tehrany, M.S., Pradhan, B., Mansor, S., Ahmad, N.: Flood susceptibility assessment using GIS-based support vector machine model with different kernel types. CATENA **125**, 91–101 (2015)
16. Varma, M.K.S., Rao, N.K.K., Raju, K.K., Varma, G.P.S.: Pixel-based classification using support vector machine classifier. In: 2016 IEEE 6th International Conference on Advanced Computing (IACC), pp. 51–55 (2016)
17. Maldonado, S., López, J.: Synchronized feature selection for support vector machines with twin hyperplanes. Knowl.-Based Syst. **132**, 119–128 (2017)
18. de Oliveira, J.F., Alencar, M.S.: Online learning early skip decision method for the HEVC Inter process using the SVM-based Pegasos algorithm. Electron. Lett. **52**(14), 1227–1229 (2016)

19. Boullé, M., Charnay, C., Lachiche, N.: A scalable robust and automatic propositionalisation approach for Bayesian classification of large mixed numerical and categorical data. Mach. Learn. **108**(2), 229–266 (2019)
20. MacNamara, A., Rabinak, C.A., Kennedy, A.E., Fitzgerald, D.A., Liberzon, I., Stein, M.B., Phan, K.L.: Emotion regulatory brain function and SSRI treatment in PTSD: neural correlates and predictors of change. Neuropsychopharmacology **41**(2), 611 (2016)
21. Yadava, M., Kumar, P., Saini, R., Roy, P.P., Dogra, D.P.: Analysis of EEG signals and its application to neuromarketing. Multimedia Tools Appl. **76**(18), 19087–19111 (2017)
22. Faust, O., Acharya, U.R., Adeli, H., Adeli, A.: Wavelet-based EEG processing for computer-aided seizure detection and epilepsy diagnosis. Seizure **26**, 56–64 (2015)
23. Gu, B., Sheng, V.S.: A robust regularization path algorithm for ν-support vector classification. IEEE Trans. Neural Netw. Learn. Syst. **28**(5), 1241–1248 (2016)

A Literature Review for Non-player Character Existence in Educational Game

Andhik Ampuh Yunanto, Darlis Herumurti, Siti Rochimah, and Siska Arifiani

Abstract Educational games have a critical role in children not only for entertainment but also for education. Unfortunately, they are not accessible, although educational games can improve student's knowledge. This paper addresses a systematic literature review on the existing research to identify and classify information about the educational game. The automatic query searches found a total of 892 papers in the digital database from 2015 until 2018, and only 87 papers selected as a quality assessment review. From 87 reviewed papers, only 18 papers fulfilled the quality assessment. The results show that the most common reason why the educational game is not accessible because it does not have a non-player character (NPC). Furthermore, the review indicates information about the game trend and suggestion about the game element for making an educational game. Besides having an educational element, we concluded that the educational game should also have NPC on the gameplay to be more interested.

Keywords Systematic literature review · Educational game · Non-player character · Game assessment

1 Introduction

In this era, the game is an entertainment application that is very popular among adults, teenagers, and children. Based on the Federation of American Scientists reports, kids aged eight to eighteen play digital games for 50 min per day on average [1]. Other research, children aged 9–16 spend 88 min per day playing the computer, and the second most common activity is playing games on the computer [2]. In

A. A. Yunanto (✉)
Informatics and Computer Engineering Department, Politeknik Elektronika Negeri Surabaya, Surabaya, Indonesia
e-mail: andhikyunanto@gmail.com

D. Herumurti · S. Rochimah · S. Arifiani
Department of Informatics, Institut Teknologi Sepuluh Nopember Surabaya, Subraya, Indonesia

E. Joelianto et al. (eds.), *Cyber Physical, Computer and Automation System*, Advances in Intelligent Systems and Computing 1291,
https://doi.org/10.1007/978-981-33-4062-6_20

the development of information technology, games have developed toward more modern technology. Several games called severe games in the marketplace have a purpose not only for entertainment but also for education. One of the serious games that often built and researched currently is about educational games that have a purpose as educational. Although educational games can be an alternative way to provide education to users [3], according to surveys, educational games are not quite accessible for users than other types of games. Other surveys also have shown that serious games or educational games not ranked in user popularity [4]. This result shows that it is necessary to find relevant information on educational games.

Artificial intelligence has an essential role for people in the future. Artificial intelligence is the realm of science that is quite popular nowadays, which implemented in some software such as automation applications, robots, and recognition systems. Besides, artificial intelligence is used as a supporting feature of interest in a game such as action games. According to the survey, action games have a high level of popularity among current users. The possible reasons are that action game has exciting features such as the existence of artificial intelligence that can provide dynamic game elements. So, it is necessary to find information on the influence of the relationship between artificial intelligence and games. Some research examples about NPC in a game are usually using reinforcement learning and natural language processing method [5]. Reinforcement learning is also one of the methods which are used to build NPC and has a better result of flexibility [6]. The other method for NPC or artificial intelligence also can use natural language processing and heuristics to answer text questions [7].

Several organizations called publishers such as the IEEE Computer Society, Elsevier, and Association for Computing Machinery (ACM) support research about game software. The organization has provided information about the methods and principles of game development in practice rather than in theory so that the knowledge gained in practice has a real impact on the usefulness of game development [8].

In general, using a game is one method to help a user to achieve goals using practice and have no risk to the environment. Besides, a game also is a software that requires a theoretical basis and literature in the design and manufacture. So, this study aims to provide information by reviewing current researches on the types and features of educational games. This review can be useful for game developers as a reference for making a good game.

The focus of this research is to conduct a review and assessment of the latest educational game research on the types, methods, techniques, procedures, and evaluations used. Moreover, the contributions in this study are summarizing the results of studies on the existence of artificial intelligence in the latest educational games in different research. Also, this study provides references for researchers who are interested. Contribution in this study is finding information about the presence or absence of artificial intelligence in the latest educational game research. So, this information can be useful to answer the problems about the existence of the educational game and to provide solutions to the development of the next educational game.

2 Systematic Literature Review Methodology

Systematic literature review, often called SLR, is a review method to identify, evaluate, and interpret the research question or topic area from all relevant literature research [5]. Primary studies are the contribution of individual research to a systematic review, and secondary studies are a form of the systematic review itself. Charters and Kitchenham [9] proposed guidelines on software engineering about SLR in the broader scope.

Serious games are software that has potential benefits such as risk-free environment, and serious games also have excellent benefits for users [8]. Based on a systematic literature review in serious games, results from research have shown empirical evidence about a user's positive effect [10]. In previous research, a literature study on the severe game about evaluations is using relevant keyword and query, which is shown in Table 1.

The query or Boolean expression for the keyword is "($A3$ OR $A1$ OR $A2$) AND ($B2$ OR ($B1$ AND ($C3$ OR $C1$ OR $C2$)))" which used to search about related research. The study is searching for related papers from 2009 until March 2015, which published on IEEE, ACM Digital Library, Springer Link, Web of Science, and Scopus. Besides, that research question is about the application domain, type of game, method, quality, procedure evaluation, and size of the population, in which all related research used as an assessment parameter in conducting SLR [8].

According to this research, our previous study also reviewed and continued about similar research questions on serious games from 2015 to 2016 but with a smaller scope [11]. Previous research, we also build an educational game using agents [12] and build an educational application using virtual reality [13]. Its scope is only reviewed paper in IEEE Explore and Science Direct. This result indicated that popular or trend serious game research is similar to previous years. Therefore, serious game development in two years has not changed. Because of this result, we should propose other perimeters for the serious game in further work.

Table 1 Keyword searching method

Code	Quantity
A1	Assessment
A2	Evaluation
A3	Validation
B1	Simulation game
B2	Serious game
C1	Training
C2	Education
C3	Teaching

3 Method

This research has the same method as literature studies in general. This research has the main objective to identify the useful information about design and method to assess the educational game. The step of the systematic literature review method is to identify, collect, classify, and analyze the relevant literature about system design and method on an educational game. Paper for the literature review is collected from the published paper in Science Direct and IEEE from 2015 until 2018 in the type of journal and conference. Detailed description of the research question (RQ), strategy, and quality assessment (QA) is explained in the next section.

3.1 Research Question

Seven research questions are conducted to achieve the objectives for the systematic literature review. The main question is about trend and method in an educational game which can support to identify all information in research. The research questions of this study are:

RQ. Is artificial intelligence used in an educational game? If yes, what method for artificial intelligence?

This RQ is the main problem for which information will be sought and analyzed. This information searched by adapting the steps of the systematic literature review. As well as this, RQ also acts as a determination of quality assessment when the author will conduct subjective judgments. Because this research topic focuses on the existence of NPCs and their methods in an educational game, the essential paper to analyze is about AI or NPC discussion.

3.2 Search Strategy

In conducting systematic literature review (SLR), strategies and methods are needed for a searching paper on the related research. The SLR method is conducted from searching until assessing selected papers. The first step is the search for related research based on the query. Queries or expressions of Boolean in this study are "Game" AND ("non-player character" OR "Artificial Intelligent" OR "Agent") AND ("education" OR "Learning"). This query is entered into the advanced search feature in Science Direct and IEEE and choose 2015 until 2018. The second step is review 1, which identifies and analyzes through the paper output. Review 1 selection is still using an advanced search query and skimming method. At this step, the research paper obtained in the paper search selected. The paper selection includes:

- Elimination of papers with unrelated titles, namely "educational game" or "learning game"
- Elimination of papers that are not conferences or journals
- Elimination of non-English language
- Elimination of the paper that discusses systematic literature review (SLR).

The results of a review one took in review 2, which analyzes through the paper. In the review phase 2, an analysis is carried out in terms of abstracts, keywords, and conclusions, as well as skimming analysis of content and images. The remaining paper will be selected by:

- Elimination of papers with abstracts that are not related to the educational game and non-player character
- Elimination of paper with keywords that are not one of the educational game or agent keywords
- Elimination of non-English language
- Elimination of papers with unique content or paper formats.

The results remaining in this review 2 will be analyzed in review 3, where this review analyzes in-depth the contents of the writing in the study. The selection made by selecting the quality of the paper is based on a list of quality assessment (QA). The QA list is explained in the next section.

3.3 Quality Assessment

Quality assessment or QA is formed based on a list of research questions. QA must contain a rating or score for answering all research questions. So, one research question will extract from two QA.

QA1. Does the paper discuss artificial intelligent or non-player character or agent?

QA2. Does the paper show a non-player character method?

The scoring schema for "Yes" is 1, and "No" is 0. Therefore, the quality assessment score of each publication is defined by the total number of "Yes."

4 Result and Discussion

Systematic literature review (SLR) research on the non-player character of educational games started from 2015 to 2018. The results of each process are shown in Fig. 1. Each result obtained from this research will be explained in the next section. Based on initial searching from 2015 until 2018, the proposed query is used to get all related research in Science Direct and IEEE.

The result of initial searching using query obtained 892 papers for educational games and non-player character keywords. The paper consists of 602 in Science

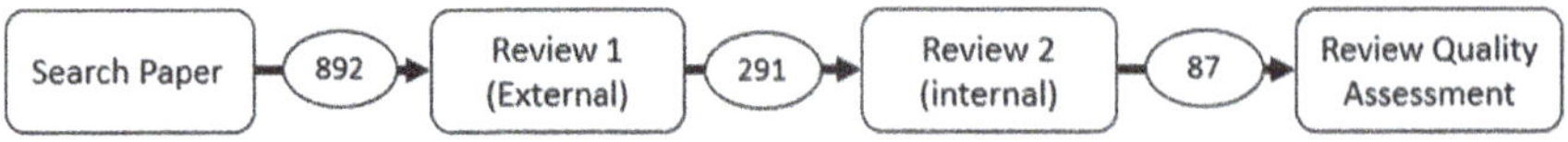

Fig. 1 Number of papers in each review process

Direct and 290 in IEEE. Next step, a paper selection was based on review 1, review 2, and following quality assessment review. The review l process received 291 related papers. The paper consists of 189 in Science Direct and 102 in IEEE. Furthermore, the review 2 processes received 87 related papers. The paper consists of 58 in Science Direct and 29 in IEEE.

4.1 *Searching and Classification Paper*

Results obtained from various points of view will be analyzed based on each step, which shows in Figs. 1 and 2. It knows that each step has a significant elimination process, from 892 to 87 papers, which means that 805 papers have been discarded and eliminated. Figure 3 shows that Science Direct has two times the amount of paper from IEEE. Figure 4 shows that there were 32 papers in 2015, 34 papers in 2016, 9 papers in 2017, and 12 papers in 2018, which took after review 2. The paper elimination from 2015 until 2018 is very high. It proves that the paper quality about

Fig. 2 Number of paper-based steps

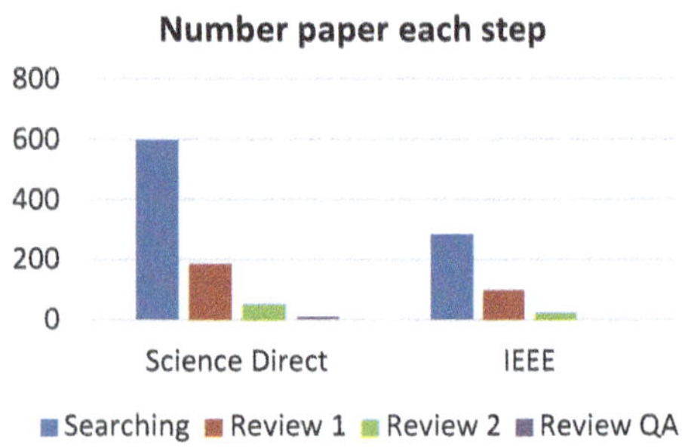

Fig. 3 Number of paper-based percentages

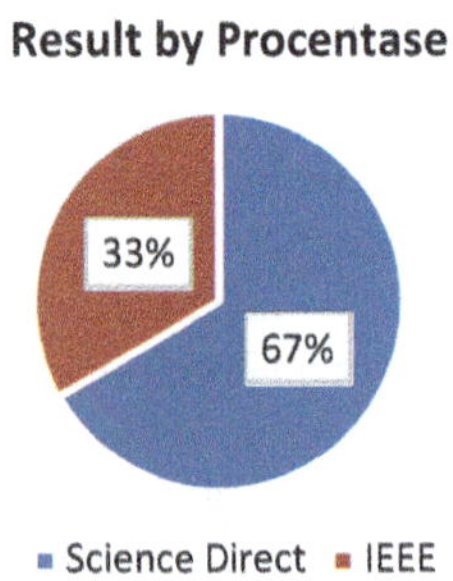

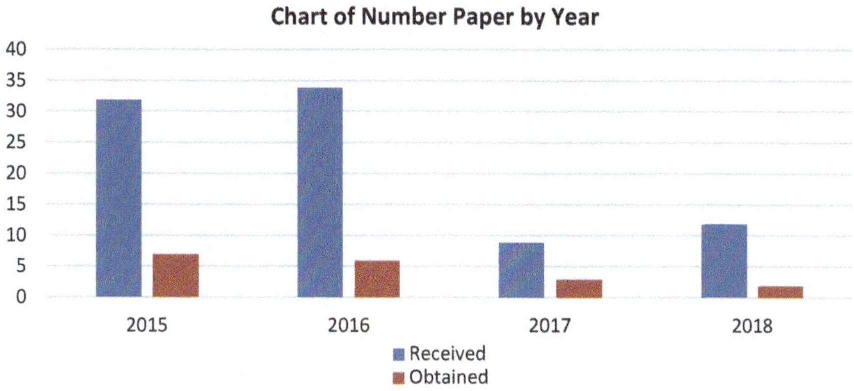

Fig. 4 Number of reviewed paper by year

the combination of NPC and educational games is still deficient. Explanation of results and analysis of each problem is explained in the next subsection.

4.2 RQ: NPC Existence and Its Method

Research question or RQ explains the classification of papers based on the analysis of non-player character and its method. Table 2 shows the list and number of papers from each NPC category. Figure 5 shows that the existing educational game research is not using NPC. In 89 researches, only 27 researches have NPC. Besides, form 27 researches, only 14 have a learning method. According to a literature review, NPC is used by the researcher but ignores an element of the educational game such as user interface, gameplay, and other designs. Also, many kinds of researches that are using NPC have just built theory and formula in many papers. Because of this reason, many researches on an educational game which has AI or NPC method are not unusual, because many kinds of research which have NPC are not a game app and ignore game element.

Furthermore, this result indicates that NPC only implements in a few types of research, which has learning methods such as reinforcement learning, probabilistic, and heuristic. However, in the end, the RQ results have provided information that

Table 2 RQ non-player character existence

NPC	Definition	Research paper	Total
NPC—no method	The game is using NPC but it has no learning method	[14–26]	13
NPC—with method	The game is using NPC and it has learning method	[27–40]	14

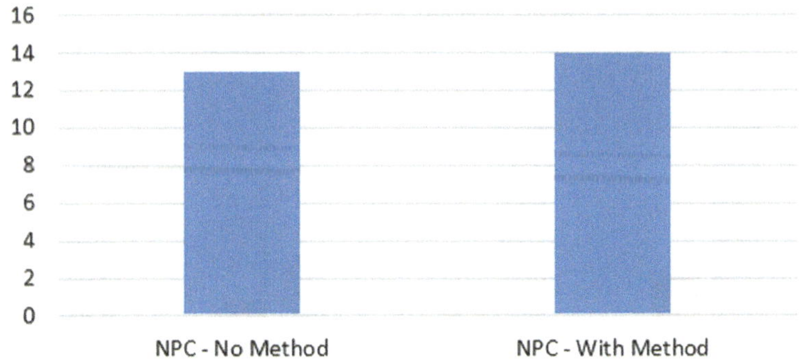

Fig. 5 Number of papers for the existence of non-player character (RQ)

the most desirable research for educational games in the last four years is about not using and implementing NPC and learning method.

Based on the results of the analysis of this review paper, there is little evidence that the current educational game is still not popular among users because it does not implement dynamic elements such as the presence of NPCs in the game. If the researchers and game developers later want to develop an educational game, it recommended implementing an NPC in it to make it more attractive to users. Besides that, by combining NPC and education, game developers can be more confident to market the game in the market place because it is a new approach game.

5 Conclusion

Information about educational game research is still needed to support contributions to the researchers. This review contributes to discussing and getting information from several studies on educational games and non-player character (NPC). This review concluded that almost educational games developed and studied did not have a non-player character and learning method. So, this information needs further investigation in further research about a combination between NPC and education. Furthermore, we assume that NPC or artificial intelligence can make an educational game more interested. The future work is to conduct a review to get information about educational games in marketplaces such as Play store (Google Play) and Windows Stores.

References

1. Scientists Federation of American: harnessing the power of video games for learning. In: Summit on Educational Games (2006)
2. Livingstone, S., Haddon, L.: EU Kids Online: Final Report. LSE: EU Kids (2009)

3. Yunanto, A.A., Herumurti, D., Kuswadayan, I., Hariadi, R.R., Rochimah, S.: Design and implementation of educational game to improve arithmetic abilities for children. In: 12th International Conference on Information and Communication Technology and System (ICTS), Surabaya, Indonesia (2019)
4. Kuswadayan, I., Herumurti, D., Hariadi, R.R., Wildianurahman, M., Yunanto, A.A., Arifiani, S.: Survival education for user on unknown islands using simulation games. In: 12th International Conference on Information and Communication Technology and System (ICTS), Surabaya, Indonesia (2019)
5. Yunanto, A.A., Herumurti, D., Rochimah, S., Kuswardayan, I.: English education game using non-player character based on natural language processing. In: The Fifth Information Systems International Conference, Surabaya, Indonesia (2019)
6. Glavin, F.G., Madden, M.G: Adaptive shooting for bots in first person shooter games using reinforcement learning. In: IEEE Transactions on Computational Intelligence and AI in Games (2015)
7. Yunanto, A. A., Herumurti, D., Kuswardayan, I.: Artificial intelligence of educational game for English learning based on similarity heuristic approach. Jurnal Sistem dan Informatika (JSI) (2018)
8. Calderon, A., Ruiz, M.: A systematic literature review on serious games evaluation: an application to software project management. Comput. Educ. 396–422 (2015)
9. Kitchenham, B., Charters, S.: Guidelines for performing systematic literature reviews in software engineering. Keele University and Durham University Joint Report (2007)
10. Connolly, T.M., Boyle, E., MacArthur, E., Hainey, T., Boyle, J.M.: A systematic literature review of empirical evidence on computer games and serious games. Comput. Educ. 661–686 (2012)
11. Yunanto, A. A., Rochimah, S.: Systematic literature review toward software evaluation in serious game. J. Informatika (2017)
12. Yunanto, A.A., Herumurti, D., Kuswadayan, I., Rochimah, S.: Intelligent system for agent in educational game using dynamic gram similarity. In: 2018 International Seminar on Application for Technology of Information and Communication, Semarang, Indonesia (2018)
13. Herumurti, D., Yuniarti, A., Rimawan, P., Yunanto, A.A.: Overcoming Glossophobia based on virtual reality and heart rate sensors. In: 2019 IEEE International Conference on Industry 4.0, Artificial Intelligence, and Communications Technology (IAICT), Bali, Indonesia (2019)
14. Safitri, A.G., Prihatmanto, A.S., Rusmin, P.H.: Design and implementation of educational game based on thematic curriculum using three layered thinking model (case study: applying number and social arithmetic in the real life). In: 4th International Conference on Interactive Digital Media (ICIDM), Bandung (2015)
15. Muntean, C.H., Andrews, J., Muntean, G.-M.: Final frontier: an educational game on solar system concepts acquisition for primary schools. In: 17th International Conference on Advanced Learning Technologies (2017)
16. Castellar, E.N., All, A., de Marez, L., Looy, J.V.: Cognitive abilities, digital games and arithmetic performance enhancement: a study comparing the effects of a math game and paper exercises. Comput. Educ. 123–133 (2015)
17. van der Graaf, J., Segers, E., Verhoeven, L.: Discovering the laws of physics with a serious game in kindergarten. Comput. Educ. 168–178 (2016)
18. Lehtonen, E., Sahlberg, H., Rovamo, E., Summala, H.: Learning game for training child bicyclists' situation awareness. Accid. Anal. Prev. (2016)
19. Tong, T., Guana, V., Jovanovic, A., Tran, F., Mozafari, G., Chignell, M., Stroulia, E.: Rapid deployment and evaluation of mobile serious games: a cognitive assessment case study. Procedia Comput. Sci. 96–103 (2015)
20. Moya, S., Tost, D., Grau, S., van Barnekow, A., Felix, E.: SKETCH'NDO: a framework for the creation of task-based serious games. J. Vis. Lang. Comput. 1–10 (2016)
21. van de Sande, E., Segers, E., Verhoeven, L.: The role of executive control in young children's serious gaming behavior. Comput. Educ. 432–441 (2015)

22. Zhanga, B., Chenc, Y., Wang, C.-Y., Liu, K.: A Chinese restaurant game for learning and decision making in cognitive radio networks. Comput. Netw. 117–134 (2015)
23. Byun, J., Loh, C.S.: Audial engagement: effects of game sound on learner engagement in digital game-based learning environments. Comput. Human Behav. 129–138 (2015)
24. Espositoa, C., Castiglionea, A., Palmieria, F., Pop, F.: Distributed strategic learning and game theoretic formulation of network embedded coding. J. Comput. Sci. 426–438 (2018)
25. Han, S., Li, X., Yan, L., Liu, Z., Guan, X.: Game-based hierarchical multi-armed bandit learning algorithm for joint channel and power allocation in underwater acoustic communication networks. Neurocomputing 166–179 (2018)
26. Süße, T.: Improvisation as a prerequisite for the dynamic interplay of production and service in PSS: insights of an organizational design principle and a game-based learning approach. In: 7th Industrial Product-Service Systems Conference—PSS, Industry Transformation for Sustainability and Business (2015)
27. Yoon, D.-M., Kim, K.-J.: Challenges and opportunities in game artificial intelligence education using angry birds. IEEE Access 2169–3536 (2015)
28. Fauzia, N., Rohendi, D., Riza, L.S.: Implementation of the cellular automata algorithm for developing an educational game. In: 2nd International Conference on Science in Information Technology (ICSITech) (2016)
29. Yang, C.: Learning methodologies for wireless big data networks: a Markovian game-theoretic perspective. Neurocomputing 431–438 (2016)
30. Bonatti, A., Hörner, J.: Learning to disagree in a game of experimentation. J. Econ. Theory 234–269 (2017)
31. Qin, Z., Khawar, F., Wan, T.: Collective game behavior learning with probabilistic graphical models. Neurocomputing 74–86 (2016)
32. Zhongjie, L., Hong-tao, L.H.: Consensus based on learning game theory with a UAV rendezvous application. Chin. J. Aeronaut. 191–199 (2015)
33. Feng, Q., Bi, W., Chen, Y., Ren, Y., Yang, D.: Cooperative game approach based on agent learning for fleet maintenance oriented to mission reliability. Comput. Ind. Eng. 221–230 (2017)
34. Cadre, H.L., Bedo, J.-S.: Dealing with uncertainty in the smart grid: a learning game approach. Comput. Netw. 15–32 (2016)
35. Zhang, X., Bao, T., Yu, T., Yang, B., Han, C.: Deep transfer Q-learning with virtual leader-follower for supply-demand Stackelberg game of smart grid. Energy 348–365 (2017)
36. Rosyida, H.A., Palmerleeb, M., Chen, K.: Deploying learning materials to game content for serious education game development: a case study. Entertainment Comput. 1–9 (2018)
37. Lahkar, R.: Equilibrium selection in the stag hunt game undergeneralized reinforcement learning. J. Econ. Behav. Organ. 63–68 (2017)
38. Kasahara, K., Nimoto, K., Takahashi, K., Inaba, M.: Interactive learning of a FALCON for a card game. In: 20th International Conference on Knowledge Based and Intelligent Information and Engineering Systems (2016)
39. Salgado, M., Clempner, J.B.: Measuring the emotional state among interacting agents: a game theory approach using reinforcement learning. Expert Syst. Appl. 266–275 (2018)
40. Quan, J., Yang, X., Wang, X.:Spatial public goods game with continuous contributions based on particle swarm optimization learning and the evolution of cooperation. Physica A 973–983 (2018)

The manufacturer's authorised representative in the EU is Springer Nature Customer Service Centre GmbH, Europaplatz 3, 69115 Heidelberg, Germany. If you have any concerns regarding our products, please contact ProductSafety@springernature.com

Printed and bound by CPI Group (UK) Ltd, Croydon, CR0 4YY
30/09/2026
02225251-0004